Exploring the Natural World: A Journey Through Fascinating Facts

A Series of Interesting and Fun Facts about Oceans, Mountains, Forests, Deserts and Weather

Jemma Stone

Copyright © 2024 Jemma Stone

All rights reserved. No portion of this book may be reproduced in any form without prior permission from the copyright owner of this book.

Table of Contents

Introduction 4
The Ocean 5
    Vastness and importance of the oceans 5
    Deep-sea creatures and marine biodiversity 13
    Currents and tides 24
Mountains 34
    Mountain Formation 34
    Ecosystems 40
    Mountains and humans 48
Forests 63
    Types of forest 63
    Ecosystems 75
    Forests and humans 84
Deserts 94
    Types of deserts 94
    Ecosystems 100
    Human survival in deserts 114
Weather 123
    Elements of weather 123
    Weather forecasting 127
    Extreme weather events and climate change 136
Thank you for reading! 153

Introduction

Have you ever looked up at a giant mountain peak or down at a massive canyon and wondered how it was formed? Have you wondered what lies in our deepest oceans, how life can survive in a scorching desert, or how extreme weather systems form and have disastrous effects? There are a great variety of ecosystems in our solar system, each very different from each other.

This book, "Exploring the Natural World: A Journey Through Fascinating Facts," invites you to discover the secrets of our planet. We will look into the ocean's dark depths, where hundreds of thousands of species lurk in the darkness. We'll go up the world's mightiest mountains, where life can be found in every corner, and high altitudes challenge even the most experienced explorers. We will look at the unique plants and animals found in lush forests, while vast deserts, sculpted by wind and sand, show their surprising resilience. Lastly, we look to the skies to discover the ever-changing weather, where environments are shaped by electrifying thunderstorms, devastating hurricane formations, and much more.

This book is divided into five chapters, each focusing on an ecosystem: Oceans, Mountains, Forests, Deserts, and Weather. Whether you're an experienced nature enthusiast or simply curious about the world around you, this journey promises to pique your curiosity and deepen your appreciation for our world's extraordinary beauty and complexity.

The Ocean

Vastness and importance of the oceans

1. **About 71% of the Earth's surface is water-covered, around 96.5% of which is held by the oceans.**
   The Pacific Ocean on its own is bigger than all of Earth's landmass combined, covering over 30% of the Earth's surface. Water makes all life on Earth possible, and it is the only planet in the solar system known to have liquid water. Life continues to exist on Earth because it is unique in that it is at the perfect distance from the Sun: not too far away that the water freezes and not too close that it gets too hot and evaporates. This area is known as the habitable zone or circumstellar habitable zone.

2. **The deepest known point on Earth is Challenger Deep, located in the Mariana Trench in the western Pacific Ocean.**
   It reaches a depth of 10,984 metres (36,037 feet), which is as deep as Mount Everest is high, plus an extra 2 kilometres! Challenger Deep was first discovered by the British ship HMS Challenger on its expedition of 1872-1876, where the crew came upon it by chance, being pushed over the undersea canyon by winds while

attempting to make landfall at a nearby island. The team took the opportunity to take measurements of the depth using a weighted rope. Despite this early exploration, it wasn't until the 20th century that technological advancements allowed for further exploration and a more detailed understanding of Challenger Deep.

3. **The oceans are a massive source of oxygen production: approximately 50-85% of the oxygen production on Earth comes from the ocean.**
Microscopic marine plants called phytoplankton undergo photosynthesis, releasing oxygen into the atmosphere. Because the numbers constantly fluctuate, it is difficult to determine the precise percentage of oxygen produced in the water. Seasons, variations in the water's nutrient load, temperature, and other variables all affect the number of plankton.

4. **Oceans are crucial in shaping the planet's climate, regulating temperatures, and supporting ecosystems.**
The vastness of the oceans is not just a geographical curiosity; it's a critical component in regulating the planet's climate. One crucial mechanism is the absorption and distribution of solar radiation. Oceans absorb vast amounts of heat from the Sun and store it in its depths. This process helps moderate temperature extremes, regulating both daily and seasonal variations. Climate patterns are also influenced by the oceans' currents, as they distribute heat around the globe.

Warm currents transfer heat from the equator towards the poles, while cold currents move in the opposite direction. This redistribution of heat affects atmospheric circulation, influencing weather patterns and helping to maintain a relatively stable climate.

5. **Oceans are essential for controlling the Earth's carbon cycle. By absorbing enormous amounts of carbon dioxide from the atmosphere, they reduce the effects of climate change.**

    Carbon dioxide from the atmosphere dissolves directly into the ocean's surface waters. The dissolved carbon dioxide is then transported and distributed throughout the ocean, where it can be stored for extended periods. Marine organisms, particularly phytoplankton and other marine plants, also play a vital role in the oceanic carbon cycle. Through photosynthesis, these organisms use sunlight to convert carbon dioxide and water into organic carbon compounds, releasing oxygen. The organic carbon may then be consumed by other marine organisms, transferring the carbon through the food web.

6. **The ocean floor is constantly moving.**

    The ocean floor undergoes constant transformation through a geological process known as plate tectonics. The solid outermost part of the Earth called the lithosphere, is divided into several large plates that float on the semi-fluid asthenosphere beneath them. These plates interact at their boundaries, leading to various geological features on the ocean floor. When tectonic plates move away from each other, it allows magma to rise from the mantle and solidify, creating new oceanic crust. This is known as a divergent boundary. When two plates move towards one another, it is called a convergent boundary. When plates collide, subduction zones can occur where one tectonic plate sinks beneath another, forming deep ocean trenches.

Additionally, transform plate boundaries occur when two plates slide past each other. As these tectonic processes unfold, they contribute to the continuous shifting and reshaping of the ocean floor. Advances in marine geophysics and technology have enabled scientists to map and study these dynamic processes.

7. **The Mid-Atlantic Ridge is the longest mountain range in the world. 90% of it lies underwater and stretches for almost 65,000 kilometres (40,390 miles).**
It extends down the centre of the Atlantic Ocean, serving as a prominent feature of the world's mid-ocean ridges. This ridge, mostly hidden beneath the ocean's surface, contains interconnected underwater mountain chains and deep rift valleys. Mid-ocean ridges occur along the divergent plate boundaries of the North American, Eurasian, South American, and African tectonic plates. As the tectonic plates separate, molten magma rises from the Earth's mantle to create new oceanic crust through volcanic activity. As new crust forms at the ridge, it pushes the existing plates apart, contributing to the continuous expansion of the Atlantic Ocean.

8. **The "Ring of Fire" in the Pacific Ocean is the name given to a belt of underwater volcanoes and earthquake sites.**

   The Ring of Fire is a horseshoe-shaped zone about 40,000 kilometres (25,000 miles) long and contains more than 450 volcanoes. This region is home to around two-thirds of the world's active and dormant volcanoes and around 90% of the world's earthquakes. The Ring of Fire lies on the convergent boundaries of different tectonic plates created by subduction, where one plate goes underneath another. When this happens, the melting of the plates produces hot magma that expands and pushes upwards through the overlying plate, erupting to the surface as a volcano.

9. **The Red Sea, which lies between Africa and Asia, was formed due to the African and Arabian tectonic plates tearing away from each other.**

   Approximately 30 million years ago, the Arabian Peninsula and the northeastern part of Africa began drifting apart due to the rifting caused by the divergence of the African and Arabian tectonic plates. This tectonic activity created a series of fractures, or rift valleys, in the Earth's crust. The rift gradually deepened and widened as the Arabian Peninsula moved eastward and the African continent moved westward. The Red Sea itself is the result of this ongoing rifting process. Over millions of years, the widening rift allowed seawater from the Indian Ocean to fill the gap, forming a new oceanic basin. This process, known as continental rifting, continues today, and the Red Sea continues to widen at a rate of a few centimetres per year.

10. **Over 80% of the world's oceans remain unexplored.**
This is mainly due to the extreme conditions in the deep ocean, such as high pressure, extremely low temperatures, and total darkness resulting in zero visibility, making deep-sea environments inhospitable for conventional human exploration. As a result, scientists rely heavily on remotely operated vehicles (ROVs), autonomous underwater vehicles (AUVs), and sophisticated sonar technologies to study these deep-sea regions. Moreover, the sheer size of the ocean poses financial and logistical difficulties, which reduces the resources available for thorough exploration. The unexplored depths hold the potential for discovering new species, understanding unique ecosystems, and unravelling valuable scientific insights, making continued exploration crucial for deepening our understanding of the ocean's hidden wonders.

11. **Point Nemo, also known as the Oceanic Pole of Inaccessibility, is located in the southern Pacific Ocean and is the farthest point from any landmass on Earth. It is also known as the "spacecraft cemetery" because over 250 spacecraft have purposefully crashed there to be disposed of.**
Located approximately 2,688 kilometres (1,670 miles) from the nearest land, Point Nemo is considered one of the most remote places on the planet. The closest

inhabited land is so far away that the closest humans to Point Nemo would often be astronauts on the International Space Station when it passes overhead. Named after the famous fictional character Captain Nemo from Jules Verne's "Twenty Thousand Leagues Under the Sea," this spot is known for its extreme isolation and vast distances from civilisation. Due to its remoteness, Point Nemo is a popular site for the controlled re-entry of spacecraft, such as satellites and space stations, as it minimises the risk of debris endangering populated areas. It is where the International Space Station is to be disposed of at the end of its life.

12. **Generally, the terms "ocean" and "sea" are used interchangeably, although there are general distinctions between them: The main difference between a sea and an ocean is their size, depth, and location.**
Oceans are typically divided into five main basins: the Atlantic, Pacific, Indian, Southern (or Antarctic), and Arctic Ocean. Oceans are deeper and larger than seas, with significant variations in temperature, salinity, and marine life. Seas, on the other hand, are smaller saltwater bodies that are partially surrounded by land. They are usually found where the land and ocean meet and are often shallower than oceans. Seas are connected to oceans and can be partially or fully enclosed by landmasses, such as the Mediterranean or Caribbean Sea.

13. **Rogue waves, also known as freak waves or monster waves, are exceptionally large and unexpected ocean waves that occur sporadically in the open sea. These waves can tower several times higher than surrounding waves, reaching heights of up to 30 metres (100 feet) or more.**

    Rogue waves pose a significant threat to ships, oil platforms, and other maritime structures due to their sudden and immense force. While the exact mechanisms behind rogue wave formation are still not fully understood, they are believed to be caused by factors such as ocean currents, atmospheric conditions, and constructive interference of smaller waves, where the crest of smaller waves align. These waves "pile up", forming one giant wave. Rogue waves can occur in any ocean and under various weather conditions, making them challenging to predict and study. Advances in satellite technology and oceanographic research have led to a better understanding of rogue waves, but their unpredictable nature continues to present risks to maritime activities. Efforts to develop early warning systems and improve ship design aim to mitigate the potential dangers posed by rogue waves in the future.

14. **At 2,300 kilometres (1,400 miles) in length, the Great Barrier Reef is so large that it can be seen from space.**

    The Great Barrier Reef, which is situated off the Australian state of Queensland, is made up of about 900 islands and nearly 3,000 individual reefs. This vibrant ecosystem is home to an astonishing diversity of marine life, including over 1,500 different species of fish, 400 different kinds of coral, and many other marine creatures like dolphins, sharks, and sea turtles. Its ecological importance, which includes providing vital habitat, sustaining fishing industries, and serving as a natural barrier against storm surges, equals its magnificent beauty. Despite facing significant threats from climate change, pollution, and coral bleaching, efforts are being made worldwide to preserve its magnificence for future generations.

15. **A23a is a colossal iceberg that broke away from the Filchner-Ronne Ice Shelf in Antarctica in 1986. It's one of the largest icebergs ever recorded, with an area of approximately 3,900 square kilometres (1,500 square miles).**
    For decades, A23a was grounded on the sea bed of the Weddell Sea until 2020, when it began on a journey through the Southern Ocean. Scientists are keenly observing its path due to the potential environmental implications of its eventual melting, including the release of substantial quantities of mineral dust. In 2024, the iceberg was confirmed to be trapped on top of a Taylor column after drifting into the Antarctic Circumpolar Current (ACC), which is a current that flows in a clockwise direction around Antarctica. The Taylor column is a massive, rotating cylinder of water created by the interaction of the ocean's currents with a submerged underwater feature, in this case, a very large bump on the ocean floor called the Pirie Bank. Scientists believe A23a could be trapped, rotating in the Taylor column for years and will slow down the iceberg's melting process significantly.

Deep-sea creatures and marine biodiversity

1. **The Turritopsis dohrnii, also known as the "immortal jellyfish," is a small jellyfish that is the only known creature capable of biological immortality.**
They have a unique ability to revert their cells to their earliest form and restart their life cycle, which usually happens when the jellyfish are exposed to environmental stress like starvation, dramatic temperature changes, or physical damage. Theoretically, this process can continue indefinitely, potentially allowing the jellyfish to live forever. However, in practice, these jellyfish can still die. In nature, most Turritopsis dohrnii are likely to succumb to predation or disease before they are able to revert to their basic form.

2. **Mantis shrimp possess one of the animal kingdom's most complex and advanced visual systems.**
They have compound eyes that can move independently. Where human eyes have three types of photoreceptor cells, shrimp's eyes contain between 12 and 16 types of photoreceptor cells, allowing them to detect polarised light and a broad spectrum of colours. This remarkable vision aids them in hunting prey and avoiding predators.

3. **Living organisms, including some crustaceans, fish, jellyfish, worms, squids and microscopic plankton, display bioluminescence, the natural production and emission of light.**
These various marine creatures harness the power of bioluminescence for purposes such as attracting mates, deterring predators, luring prey, communication, and mimicry. The process occurs through the chemical reaction between luciferin and oxygen, facilitated by enzymes. In order to hide from predators or captivate prey, several species of jellyfish use bioluminescence to produce an ethereal glow.

4. **Octopuses are incredibly intelligent creatures with problem-solving abilities and complex learning behaviours.**
They have large brains relative to their body size. Research has shown that octopuses are skilled problem-solvers, capable of learning and remembering tasks for extended periods of time. Their ability to navigate through mazes, open jars to obtain food, and imitate the actions of other animals are all clear indications of their intelligence. Octopuses also demonstrate remarkable creativity, using clever strategies to escape predators and capture prey. Their mastery of camouflage, skilful use of tools, and manipulation of objects show their intelligence and problem-solving abilities. Octopuses are also lively and inquisitive, engaging in exploratory behaviours that suggest some level of consciousness and awareness. Their intelligence has led scientists to explore the possibility of extraterrestrial-like intelligence on Earth.

5. **Deep-sea gigantism is a phenomenon that many deep-sea creatures exhibit, where species grow significantly larger than their shallow-water relatives.**
   The colossal squid, which can grow as long as 10 metres, and the giant isopod, the largest of which has been measured at 50 centimetres compared with the usual 5 centimetres, are two examples. Although the reasons behind this gigantism are still not fully understood, it is thought to be related to reduced predation pressure and limited food resources, where a larger body size could be beneficial because of the improved ability to look for food over greater distances. It is challenging to study this topic due to the inaccessibility of the deep ocean.

6. **The electric eel, found in the Amazon and Orinoco River basins, can generate electric shocks for hunting and self-defence.**
   Electric eels have specialised cells called electrocytes that can produce electric charges. Because they have very poor eyesight, they use these charges to navigate and locate prey. They also use electric signals to communicate with other electric eels and even generate powerful shocks of up to 600 volts to deter predators or capture prey. This is enough to incapacitate a human or even kill them if shocked multiple times, as it can cause respiratory or heart failure. People can also die from drowning after being incapacitated by an electric eel. Despite their name and appearance, electric eels are more closely related to fish, like catfish, than eels.

7. **Blue whales, the largest animals on Earth, communicate with each other using loud, low-frequency songs that can travel across vast distances in the ocean.**

    These complex and hauntingly beautiful songs are considered crucial in mate attraction and territory establishment. Blue whales don't depend as much on good vision as land animals do due to the extreme darkness of the deep ocean. Similarly, they don't require a keen sense of smell, as scents spread more slowly in water than in air. Instead, sound becomes an essential sensory tool in the oceans, where it travels faster than in the air. Therefore, blue whales heavily rely on sound to communicate and find food sources in the wide-open ocean. Blue whales are known to have unique regional dialects, and their vocalisations provide valuable insights into their behaviour and social structure.

8. **Deep ocean hydrothermal vents have the ability to support life in a place where sunlight can't reach.**

    Discovered in 1977, hydrothermal vents are openings in the seafloor where water seeps into the ocean floor, becomes superheated from Earth's internal heat, and then bursts back into the ocean. Hydrothermal vents' ability to support life without sunlight—the energy source that forms the foundation of every other food web on the planet—makes them especially fascinating. The superheated water from these vents carries a rich mixture of minerals and chemicals, providing an energy source for specialised microorganisms such as bacteria and archaea. These microorganisms form the foundation of a food chain that supports a variety of creatures, including giant tube worms, clams, and shrimp.

9. **Parrotfish have a unique feeding mechanism where they scrape algae and living coral off rocks using their teeth.**
Their teeth are fused together and arranged on the outside of their jaw, forming a hard, beak-like structure. Mostly found in coral reefs, they grind up the material with their teeth, digest the edible portions and then excrete the remaining material as sand, helping to create sandy beaches or small islands. They can potentially contribute to coral reef erosion, but they also aid in preventing the overgrowth of algae on coral reefs. Interestingly, parrotfish also form a mucous cocoon around themselves at night before sleeping, which serves as a barrier against parasites and predators.

10. **The anglerfish hunt using a bioluminescent lure, which lies at the end of a long filament extending from the middle of their heads.**
This glowing bait, resembling a tiny lantern in the pitch-black abyss, serves two purposes for the anglerfish. Its ethereal glow attracts prey and mesmerises them, making them easier to hunt. As unsuspecting prey approach to investigate the lure, the anglerfish quickly takes advantage, ambushing and engulfing its victim in its enormous mouth. Thanks to this innovative and effective technique, anglerfish can survive in the harsh conditions of the deep ocean, where food is limited and visibility is almost non-existent.

11. **The narwhal, sometimes referred to as the "unicorn of the sea," has a long, spiral tooth that protrudes from the left side of its upper jaw and through the lip. It is capable of reaching up to ten feet in length.**
Contrary to popular belief, this elongated tooth, or tusk, is not primarily used for hunting. Instead, it serves as a sensory organ with millions of nerve endings, detecting changes in water temperature, salinity, and pressure. This helps narwhals locate prey and navigate through their Arctic habitat.

12. **The archerfish have a unique predation technique where they "shoot down" their prey to knock them into the water.**
Found in mangrove swamps and estuaries, archerfish catch prey by using specialised muscles to form a tube with its mouth, creating a focused stream of water. With incredibly precise aim, the archerfish spits a jet of water at insects above the water's surface, knocking them into the water, where the fish can easily capture and consume them.

13. **Starfish use their tube feet, located on the underside of their arms, to create suction and pry open the shells of bivalve molluscs to feed.**
Once the shell is slightly ajar, the starfish everts its stomach through its mouth and into the shell, secreting digestive enzymes to break down the soft tissues into a liquid form. The starfish then retracts its stomach, filled with the liquefied meal, back into its central disc. This

efficient process allows starfish to consume a variety of prey, including clams and mussels.

14. **Orcas, also known as killer whales, demonstrate a feeding strategy known as wave-washing. In this strategy, they create waves to knock prey into the water.**
When hunting seals or sea lions on floating ice sheets, orcas create waves with their tails to wash the prey into the water. This behaviour showcases their high-level intelligence and collaboration within the pod, as different members take on specific roles in the coordinated effort to secure a meal.

15. **Humpback whales engage in a cooperative and intricate feeding behaviour called bubble net feeding, creating "nets" of bubbles to trap schools of fish.**
Groups of up to 20 whales work together to encircle schools of fish. As the group circles, they exhale bubbles in a spiralling pattern, forming a net that traps the fish. The whales then swim through the centre of the bubble net with open mouths, swallowing large amounts of water and fish. This coordinated effort allows humpback whales to feed on schools of fish efficiently.

16. **During the late 19th and early 20th centuries, a dolphin named Pelorus Jack became famous for guiding ships through the waters of Cook Strait in New Zealand.**
Believed to have been born around 1888, the solitary dolphin, identified through photographs as a Risso's

dolphin, became well-known for guiding ships safely through treacherous waters, often swimming alongside and entertaining passengers with acrobatic displays. Pelorus Jack's appearances were documented in numerous ship logs and newspapers of the time. The New Zealand government eventually declared Pelorus Jack a protected marine mammal in 1904. His legacy endures as a symbol of the unique relationships that can develop between humans and wild animals and the mysterious intelligence of dolphins. The dolphin disappeared in April 1912, although the cause of his death was never confirmed.

17. **Various larger aquatic species gather at locations known as "cleaning stations" to be cleaned by smaller animals.**

    These stations, typically found on coral reefs, are created and maintained by cleaner fish, such as cleaner wrasses. Larger marine animals, such as sea turtles, reef fish, and even sharks, visit these stations to undergo a cleaning process. The cleaner fish thoroughly remove parasites, dead skin, and other debris from the bodies of the bigger animals, even swimming into the mouth or gills of the client. This cleaning behaviour not only promotes the health and hygiene of the larger marine animals but also creates a mutually beneficial relationship. The cleaner fish get a nutritious meal from the parasites, while the clients receive a service that aids in preventing diseases and parasites.

18. **The blobfish uses a minimal-energy strategy for capturing prey, patiently waiting for small invertebrates to drift into its path.**

The blobfish, residing in the deep ocean depths, has evolved a feeding strategy that aligns with its buoyant lifestyle. With a density slightly less than water, the blobfish hovers just above the seafloor without expending much energy, which is critical to their survival. This energy-efficient approach allows the blobfish to conserve its resources in the low-food environments of the deep sea. The blobfish's diet primarily consists of small invertebrates that float by in the ocean currents. Its downturned mouth allows it to scoop up prey effortlessly, making the most of its habitat and ensuring its survival in the dark and pressurised depths where it resides.

19. **Coral reefs take up less than 1% of the space in the ocean, but provide a home for at least 25% of marine life.**

Marine animals have many uses for the coral reef environment. The complex structures of coral reefs create diverse microhabitats, offering shelter from predators, breeding grounds, and feeding areas for countless species.

20. **Coral reefs have many natural threats that can cause their death, such as disease or invasive species.**

Coral reefs face natural challenges that can impact their health and resilience. One significant natural threat is predation by crown-of-thorns starfish, which are coral predators capable of consuming massive amounts of coral tissue. Outbreaks of crown-of-thorns starfish can lead to widespread coral death and have been known to cause considerable damage to reef ecosystems. Additionally, certain coral diseases, such as white plague and black band disease, are caused by naturally occurring pathogens, which can also cause widespread coral death.

21. **Human activities such as overfishing, pollution, blast fishing, and coral over-mining pose severe threats to coral reefs.**

    Overfishing, driven by an increasing demand for seafood, disrupts the delicate balance within coral reef ecosystems by depleting key species. Pollution from coastal runoff introduces sediments, nutrients, and harmful chemicals into reef environments, leading to degraded water quality. Destructive fishing practices, such as blast fishing by using explosives to catch big groups of fish easily, further exacerbate the problem by directly damaging coral structures and harming non-target organisms. Moreover, the global trade of coral and coral products for ornamental purposes, souvenirs, and the aquarium trade puts additional pressure on these vulnerable ecosystems.

22. **Coral bleaching is the process by which stressed coral expels the algae from their living tissues, causing them to lose their vivid colour and turn them entirely white.**

    These algae, known as zooxanthellae, contribute to the vibrant colours of the reef and supply corals with vital nutrients through photosynthesis. However, when corals experience environmental stressors such as elevated sea temperatures, pollution, or alterations in water quality, they expel the algae, which gives the corals a bleached appearance. While the loss of algae doesn't indicate that a coral is dead, it does deprive it of an essential energy source, decrease its resilience, and make them more susceptible to diseases. Extended or severe coral bleaching can result in widespread coral death, a serious threat to the complex biodiversity and ecological balance of coral reef ecosystems.

23. **Sea cucumbers have a unique defence mechanism: they can expel their internal organs, which are later regenerated. This behaviour helps them escape from predators.**

    When threatened by predators, sea cucumbers employ several strategies to protect themselves. One of the most remarkable methods is their ability to expel their internal organs through a process called evisceration. Sea cucumbers expel their digestive tract and respiratory organs by forcibly contracting their muscles and rupturing their body wall, which can distract or deter predators. While this may seem extreme, sea cucumbers have the remarkable ability to regenerate lost organs over time. Additionally, some species of sea cucumbers can release toxic chemicals into the surrounding water as a defence mechanism. These chemicals can be harmful or distasteful to predators, deterring them from further attacks.

Currents and tides

1. **The moon's and, to a lesser extent, the Sun's gravitational pull cause the oceans of Earth to rise and fall regularly, causing tides.**
   The primary influence comes from the moon, whose gravitational pull creates bulges of water on the side facing it and on the opposite side of the Earth, creating the high tides. As the Earth rotates within these bulges, most coastal areas experience two high tides and two low tides daily. The Sun also plays a role, though its effect is less pronounced than the moon's. Spring tides are when the Sun and Moon's gravitational pull align during full and new moons, causing higher high tides and lower low tides. Conversely, neap tides, which have lower high tides and higher low tides, are caused by the Sun lessening the moon's gravitational pull on Earth during the moon's first and third quarters, when the moon and the Sun are at right angles to one another.

2. **It takes 12 hours and 25 minutes for the tide to go out and then back in again from high tide.**
   A "lunar day", the time it takes the moon to travel around the Earth, is slightly longer than a typical solar day, at 24 hours and 50 minutes. Most coastal areas will have two high tides and two low tides in a 24-hour and 50-minute period, which is why high and low tides occur at different times every day.

3. **The Coriolis effect is a phenomenon in physics resulting from the Earth's rotation on its axis. This rotation causes moving objects on the**

Earth's surface to be deflected. This means they don't travel in a straight line, even if they are moving in a straight path relative to their own perspective.

Named after the French mathematician Gaspard-Gustave de Coriolis, who first described it in the early 19th century, this effect influences the trajectory of winds, ocean currents, and objects in motion, such as projectiles. In the Northern Hemisphere, moving objects tend to be deflected to the right, while in the Southern Hemisphere, they deflect to the left. The Coriolis effect arises due to the conservation of momentum as the Earth rotates beneath objects in motion. Understanding the Coriolis effect is crucial in various fields, such as meteorology, oceanography, and even long-range artillery targeting, as it significantly influences the paths of moving objects over large distances on Earth.

4. **Saltstraumen, located in Norway near the Arctic Circle, has one of the strongest tidal currents on the planet. As the tide rises and falls, up to 400 million cubic metres of water move through a 150-metre (490-foot) wide strait about every six hours.**

Saltstraumen is a stretch of water that is very narrow and connects the sea to a fjord, through which powerful tidal currents are demonstrated. The tidal currents in Saltstraumen can reach speeds of up to 22 knots, creating maelstroms and powerful whirlpools. The changing tides cause massive volumes of water to surge through the strait, leading to rough sea conditions.

5. **The Earth's oceans are interconnected by a system known as the Global Conveyor Belt, a massive circulation pattern driven by differences in temperature and salinity (saltiness).**
Surface and deep ocean currents make up the global conveyor belt, which circulates the world once every 1,000 years. Warm surface currents move toward the poles from the equator, where they cool, become denser, and sink to the deep ocean. Then, cold, deep ocean currents carry the denser water away from the poles towards the equator, creating a continuous flow that redistributes heat around the globe and has a role in climate regulation.

6. **Climate change could cause the Global Conveyor Belt to slow or even be stopped.**
Climate change leads to increases in ocean temperatures, evaporation of seawater, and the melting of glaciers and polar ice caps, which may cause warm freshwater to rise onto the ocean's surface. This influx of freshwater disrupts the normal density and salinity patterns that drive the global conveyor belt. The melting ice dilutes the seawater, making it less dense and preventing it from sinking into deeper ocean layers. This disruption can potentially weaken or even shut down parts of the global conveyor belt, leading to altered heat distribution. Such changes can profoundly affect marine ecosystems, weather patterns, and global climate regulation.

7. **Gyres are large, circular ocean currents that are known for trapping and accumulating marine debris, leading to the formation of ocean garbage patches.**
These currents are primarily driven by the combination of global wind patterns, the Earth's rotation, and the distribution of continents. There are several major gyres in the world's oceans, such as the North Atlantic Gyre and the South Pacific Gyre; they flow clockwise in the Northern Hemisphere and counterclockwise in the Southern Hemisphere. The dynamics of gyres contribute significantly to the planet's oceanic circulation, affecting everything from temperature patterns to the distribution of marine life. The circular flow of these gyres means that human-created waste, like plastics dumped into the sea, can get trapped, forming large patches of waste.

8. **The largest garbage patch in the world is the Great Pacific garbage patch.**
The Great Pacific garbage patch is a large, floating collection of marine debris, primarily composed of plastic, in the central North Pacific Ocean. It is a collection of dispersed microplastics and larger debris, rather than a solid mass, that stretches across an area estimated to be larger than some countries. Because plastic pollution can entangle and harm animals, as well as introduce chemicals into the food chain through the breakdown of plastics into microplastics, it seriously endangers marine life. Cleanup efforts face significant challenges due to the patch's immense size and the fragmented nature of the debris.

9. **Intertidal ecology is the study of ecosystems between the high and low tide marks, where land experiences regular submersion and exposure.**
Also known as the intertidal zone, this dynamic environment undergoes constant temperature, salinity, and moisture fluctuations due to tidal cycles. Organisms living in the intertidal zone, such as barnacles, molluscs, seaweeds, and crustaceans, have evolved unique adaptations to survive both underwater at high tide and exposed to air during low tide. These adaptations may include specialised physiological processes, protective shells, or the ability to anchor themselves securely to rocks. Intertidal ecosystems serve as nurseries for many marine species and provide feeding grounds for birds.

10. **The tides on either side of the Cook Strait, which is located between the North and South Islands of New Zealand, are out of phase with each other. High tide on one end of the strait occurs at almost the same time as low tide on the opposite side.**
The Cook Strait is a 22-kilometre (14-mile) long, narrow, and tumultuous stretch of water that serves as the main connection between the Tasman Sea to the west and the Pacific Ocean to the east. High tide occurs on the Pacific Ocean side of the strait five hours before high tide on the Tasman Sea side. The difference in water levels causes fast currents reaching up to 2.5 metres per second.

11. **The Sargasso Sea, located in the North Atlantic Ocean, is a unique region known for its vast floating mats of Sargassum seaweed. Unlike traditional seas, it is defined not by land borders but by ocean currents.**

    The Sargasso Sea is surrounded by four different currents which trap seaweed within its circular flow:
    - The North Atlantic current to the north
    - The Canary Current to the east
    - The North Atlantic Equatorial current to the south
    - The Gulf Stream to the west

    The Sargassum provides essential habitat for marine life, including fish, crabs, and young sea turtles. The Sargasso Sea has a vibrant ecosystem despite its supposedly barren appearance. The Sargassum seaweed has distinctive characteristics, such as gas-filled bladders that allow it to float, creating a habitat that supports a complex food web. The Sargasso Sea is crucial in global ocean processes and acts as a spawning ground for various species. However, it faces threats such as pollution; the Sargasso Sea accumulates a lot of plastic waste due to the currents surrounding it.

12. **The horse latitudes are subtropical regions located about 30 degrees north and south of the equator, known for their calm winds and little to no rain.**

    These latitudes are characterised by high atmospheric pressure and descending air masses, which cause winds to diverge and either flow towards the poles (known as the prevailing westerlies) or the equator (known as the trade winds). The descending air in the horse latitudes creates a zone of weak and inconsistent winds, making it challenging for sailing ships to navigate efficiently. It is thought that the term "horse latitudes" originated from the historical practice of European sailors throwing horses overboard to conserve water and reduce the load when their immobility in these regions caused delays to voyages.

13. **The El Niño and La Niña phenomena are climate events associated with the periodic warming (El Niño) and cooling (La Niña) of surface waters in the central and eastern Pacific Ocean.**
El Niño brings warmer than average sea surface temperatures and weakens trade winds that, in normal conditions, push warm surface waters towards Asia. During an El Niño event, it moves the warm water eastwards towards South America and reduces the upwelling of cold water off the coast. This often influences wetter winters in North America and drier conditions in the northern US and Canada. La Niña, the opposite phase, strengthens trade winds, which pushes warm water back towards Asia in the west and chills the eastern Pacific, leading to colder and drier winters in North America and increased rainfall in Southeast Asia. These cycles typically last 3-7 years and impact global weather patterns, affecting food production, wildfire risk, water resources and fisheries. Understanding El Niño and La Niña helps us predict seasonal climate trends and prepare for potential extremes.

14. **The Bay of Fundy is a bay that lies between two Canadian provinces: New Brunswick and Nova Scotia. It has the highest tidal range (difference in height between high and low tide) in the world; it can be as much as 16 metres (52 feet), whereas the typical tidal range worldwide is 1 metre (3 feet).**

    This massive tidal range is due to a phenomenon called tidal resonance. This effect occurs when the natural period of the bay's water sloshing back and forth (approximately 12 hours) occurs at almost the same frequency as the tide coming in and going out (approximately 12 hours and 25 minutes). In other words, the time taken for a wave to move from the mouth of the bay to the shore and back to the mouth is almost the exact same amount of time for the tide to go from high to low or vice versa. This synchronisation acts like a water pump, amplifying the tidal surge and forcing enormous volumes of water into the funnel-shaped bay, resulting in the highest tides on Earth.

15. **Tidal bores are phenomena that occur where estuaries and rivers flow into the ocean. The incoming tide produces a sudden and sometimes dramatic surge of water against the river's current.**

    This surge produces a visible and audible tidal bore, characterised by a rapidly rising wall of water that travels upstream. Places with more tidal bores are areas with funnel-shaped estuaries or narrowing river mouths, facilitating the amplification of tidal forces. The Qiantang River in China is well known for having one of the world's biggest tidal bores, with waves reaching up to 9 metres (30 feet). In addition to being a remarkable natural sight, tidal bores provide challenges and opportunities for communities along affected rivers. They can impact cultural events and celebrations, recreational activities, and navigation in areas where tidal bores occur regularly.

16. **Seawater, as an electrical conductor, can influence Earth's magnetic field.**
As seawater contains ions, mainly dissolved salts like sodium and chloride, it becomes a conductor of electricity. When seawater moves, such as in ocean currents or tidal flows, it cuts through Earth's magnetic field lines, inducing electric currents within the water. While the overall effect is relatively small, the oceans' vastness and constant movement contribute to Earth's magnetic environment. Variations in magnetic intensity caused by ocean currents have historically influenced compass readings and navigation, posing challenges for sailors and navigators.

17. **Harnessing energy from tidal currents is a promising renewable energy strategy that involves capturing the kinetic energy generated by the movement of tides.**
Tidal energy systems typically utilise underwater turbines strategically positioned in areas with strong tidal currents, including narrow channels or coastal passages. These turbines rotate due to the flow of tidal currents, which generate electricity. Tidal energy is appealing for its predictability and consistency, as tidal

cycles are well-known and can be predicted accurately. However, challenges include the hostile marine environment, potential effects on aquatic ecosystems, and expensive installation costs. Current tidal energy research and ongoing technological advancements aim to overcome these challenges, which means tidal power could one day contribute to global renewable energy.

Mountains

Mountain Formation

1. **Mountains are formed in one of three ways: tectonic forces, volcanic activity, or erosion.**

   Tectonic forces - tectonic plates are massive sections of the Earth's crust which move very slowly due to the plates lying on top of molten rock. When two plates collide, immense pressure builds, causing the Earth's crust to crumple and fold, giving rise to mountains. This process, known as orogenesis, is a slow geological phenomenon that unfolds over millions of years. Examples of how tectonic pressures shape some of the tallest peaks in the world are the Himalayas, which were created when the Indian and Eurasian tectonic plates collided.

   Volcanic activity - mountains born from volcanic origins result from molten rock, or magma, surging from the Earth's mantle to the surface. As magma erupts, it cools and solidifies, layering upon itself and gradually building the framework of a mountain. Some of the world's iconic peaks, such as Mount Fuji in Japan or Mount Kilimanjaro in Tanzania, are formed as a result of this process. Conical formations and rough terrain are common characteristics of volcanic mountains, which demonstrate the interaction between the Earth's crust and the extreme heat underneath. These types of mountains are formed in part by the continuous eruption and buildup of volcanic debris.

Erosion - although often associated with powerful geological forces, erosion over long periods of time is also responsible for mountain's unique formations. Erosion gradually wears away rock and soil and is crucial in shaping mountainous landscapes. As rivers carve through valleys and wind whittles away at the exposed rock, mountains transform. Over time, the relentless forces of weathering break down the rock into sediment, which is then transported by rivers and wind, leaving behind the distinct features we recognise as mountains.

2. **The Himalayas, a mountain range in Asia, are home to the world's highest peaks, including Mount Everest.**
Mount Everest stands at 8,848 metres (29,032 feet) and is the highest point on Earth. The Himalayas span five countries: India, Nepal, Bhutan, China, and Pakistan, and have over 100 peaks which are higher than 7,200 metres (23,600 feet) above sea level.

3. **While Mount Everest's summit is the highest point above sea level, it isn't actually the tallest mountain on Earth. Mauna Kea, located in Hawaii, is taller when it is measured from base to summit.**

The peak of Mauna Kea is 4,203 metres (13,803 feet) above sea level, but its base is at the ocean floor, making its total height over 10,210 metres (33,500 feet).

4. **The longest continental mountain range is the Andes, which stretches along South America's western border. It spans about 7,000 kilometres (4,300 miles).**
It goes through seven countries: Venezuela, Colombia, Ecuador, Peru, Bolivia, Chile, and Argentina. This formidable mountain range results from the Nazca Plate sliding beneath the South American Plate. The Andes showcase unparalleled geographical diversity, featuring towering peaks, deep valleys, and high-altitude plateaus. The Andes have been crucial in shaping the climate and ecosystems of the region, influencing the formation of the Amazon rainforest and the Atacama Desert. Rich in mineral resources, the Andes have historically been a source of valuable minerals like copper and silver. The region's cultural heritage is also deeply intertwined with the mountains, as they were sacred to many indigenous civilisations, including the Inca Empire.

5. **The Appalachian Mountains in North America are one of the oldest mountain ranges on Earth. They were formed approximately 480 million years ago.**
The Appalachian Mountains stretch approximately 1,500 miles from Alabama in the U.S. to Newfoundland in Canada. They are well-known for their breathtaking scenery and rich biodiversity and are home to various

habitats, from high-altitude spruce-fir stands to hardwood forests. This mountain range includes the Great Smoky Mountains, which are a UNESCO World Heritage Site and a biodiversity hotspot. Notable for their role in shaping early American history, the Appalachian Mountains served as a natural barrier to westward expansion, leading to the creation of the Appalachian Trail. This 2,200-mile-long hiking trail runs from Georgia to Maine.

6. **Mount St. Helens in Washington, USA, erupted in 1980 and was the deadliest and most economically devastating volcanic event in American history.**

    Prior to the eruption, Mount St. Helens stood at an elevation of 2,950 metres (9,677 feet). The eruption, triggered by a massive landslide and subsequent release of pressurised magma, resulted in the mountain losing about 396 metres (1,300 feet) of its height. Ash from the eruption spread across eleven U.S. states, causing severe damage to infrastructure, extensive changes to the surrounding landscape, and the tragic loss of lives. The blast created a horseshoe-shaped crater on the north side of the mountain. It caused the surrounding forests to flatten, leaving behind a barren landscape. Despite the devastation, Mount St. Helens has become a natural laboratory for the study of ecological recovery, with life gradually returning to the region. The mountain is now part of the Mount St. Helens National Volcanic Monument, attracting scientists, tourists, and outdoor enthusiasts interested in witnessing the ongoing recovery process and the geological forces that shape our planet.

7. **The iconic Matterhorn, a mountain located on the border between Switzerland and Italy, is one of the most photographed mountains in the world. It is known for its pyramid shape, which appears on Toblerone chocolate bars.**

   The Matterhorn, an iconic mountain that is part of the Alps, is one of the most recognisable peaks in the world. Towering at 4,478 metres (14,692 feet), it has a distinctive pyramid shape and sharp, jagged ridges. The mountain's four faces point toward the compass directions, and climbing routes on each face pose significant challenges, contributing to its reputation as one of the deadliest peaks in the Alps. Its stunning beauty attracts climbers, hikers, and tourists, making it one of the most photographed and sought-after destinations in the Alps.

8. **The tallest volcano in our solar system is called Olympus Mons and is situated on the surface of Mars. It stands at a height of 21,900 metres (72,000 feet) from base to summit.**

   At two and a half times the height of Mount Everest above sea level, Olympus Mons towers over its terrestrial equivalent. Mars lacks such plate tectonics, unlike Earth, where tectonic plates continually shift and create new geological features. Instead, Olympus Mons was formed over billions of years as magma welled up through a single stationary hot spot on the Martian surface, leading to the gradual accumulation of lava layers that shaped this immense shield volcano. As a result, the mountain's appearance is low and squat, with an average slope of just 5 percent. Olympus Mons last erupted approximately 25 million years ago.

9. **Despite its small size, Io, one of Jupiter's moons, is the most volcanically active body in the solar system, with hundreds of active volcanoes dotting its surface. Volcanic eruptions on Io can be so violent that they can throw columns of gas and dust far into space.**

   Being so small, Io should have lost all of its inner heat to space and therefore shouldn't have any active volcanoes on its surface. Its volcanic activity is due to the gravitational pull from Jupiter as it orbits around it, as well as the perfectly timed gravitational pulls from two other moons orbiting around Jupiter, Europa and Ganymede. This causes Io to be constantly stretched and squeezed, which creates the heat and energy that creates incredible volcanic eruptions.

10. **Mons Huygens is the tallest mountain in the Moon's Montes Apenninus mountain range, rising to about 5.5 kilometres (18,000 feet). Mons Huygens, along with the entire Montes Apenninus range, was formed roughly 3.9 billion years ago by the impact of a colossal asteroid or comet that created the Mare Imbrium, one of the Moon's large dark plains.**

    The impact, one of the largest in the Moon's history, created the Imbrium Basin. This vast, circular depression would later fill with lava, forming the dark plains known as Mare Imbrium. Massive amounts of lunar crust and rock were then pushed up to create towering mountain ranges along the basin's rim.

11. **Seamounts are underwater mountains rising from the ocean floor, typically with pointed or peaked summits that do not reach the water's surface and, therefore, do not form an island.**

    These geological features are abundant across the world's oceans, often forming in chains or clusters. Seamounts can vary significantly in size, with some towering thousands of metres above the seabed. They play crucial roles in marine ecosystems, supporting various marine life, including corals, sponges, crustaceans and fish. Additionally, seamounts can influence ocean currents and nutrient circulation, making them significant for oceanographic studies and understanding global climate patterns. These underwater mountains also hold scientific interest for their potential mineral and biological resources. There are thousands of seamounts in the world's oceans, however not all have been mapped. In 2005, a submarine called USS San Francisco struck a seamount while going at full speed in the Pacific Ocean, severely injuring most of her crew and killing one, showing the possible dangers of these uncharted seamounts.

Ecosystems

1. **A geographical and ecological phenomenon known as "altitudinal zonation" describes the**

distinct horizontal bands of vegetation, climate, and ecosystems that can be found on a mountainside based on elevation.

Different ecological zones arise due to temperature, precipitation, and air pressure changes that occur when climbing a mountain. The most common altitudinal zonation includes the following zones:

- The tropical zone at the mountain's base, containing lush forests.
- The subtropical zone with mixed forests.
- The temperate zone, often featuring coniferous or deciduous forests.
- The subalpine zone, with smaller trees and shrubs.
- The alpine zone, characterised by meadows, rocky terrain, and, at higher elevations, permanent snow and ice.

Each zone supports distinct plant and animal species that are adapted to the particular environmental conditions of their separate elevations and, therefore, plays a critical role in shaping biodiversity.

2. **Because the environments above and below a specific altitudinal zone are hostile and limit organisms' ability to move about or disperse, certain plants and animals found in altitudinal zones have a tendency to become isolated. The term "sky islands" refers to these isolated biological systems.**

Sky islands are unique and isolated mountain ecosystems distinguished by their altitude and the lowland environments that surround them. These "islands" of elevated terrain are often dispersed across lower elevations, resembling isolated peaks or mountain ranges surrounded by lower, warmer valleys. Sky islands are essential to biodiversity because of the diverse ecological zones created by their varying heights. There may be significant levels of endemism - the condition in which a species is limited to a single, defined geographic area - among the species that live on these solitary peaks, as they may be specifically adapted to withstand the unique conditions of their higher habitats. "Sky islands" refers to the concept that these mountainous areas function as biological islands in the sky, supporting unique ecosystems amidst the surrounding lower-elevation environments.

3. **The endangered mountain gorillas inhabit the volcanic slopes of the Virunga Mountains in Central Africa.**
The Virunga Mountains, a volcanic range spanning the borders of Rwanda, Uganda, and the Democratic Republic of the Congo, are home to the critically endangered mountain gorillas. Mountain gorillas are distinguishable by their thick fur, robust build, and, in the case of males, a distinctive silver stripe on their back. These gorillas faced severe threats from habitat loss, poaching, and political instability. Conservation efforts, such as ecotourism and anti-poaching, have somewhat increased their population in recent years, making the Virunga Mountains a symbol of successful conservation strategies and a sanctuary for the survival of these remarkable and endangered creatures.

4. **Puncak Jaya, a mountain peak in Oceania, has glaciers on its slopes, making it one of the few places where glaciers exist in tropical regions.**
Despite its tropical location near the equator, Puncak Jaya has glaciers, making it one of the few equatorial mountains with these icy formations. The most notable of these glaciers is the Carstensz Glacier, which clings to the mountain's upper slopes. Puncak Jaya was named "Carstensz Pyramid" after Dutch explorer Jan Carstensz, the first European to sight the glaciers on the mountain's peak on a rare clear day in 1623. Other Europeans didn't believe this sighting, which wasn't confirmed for another two decades.

5. **Cushion plants grow in high-altitude mountainous environments; they form dense, low-lying mats of tightly packed stems of different plants that resemble cushions.**
These cushion plants are designed to endure the harsh climate on elevated terrains. The compact structure is an effective defence against strong winds, providing insulation and reducing heat loss. Notably, cushion plants can withstand freezing temperatures, which are common in mountainous environments. Their tiny, leathery leaves help conserve water by reducing water loss through evaporation.

6. **The scarcity of plant pollinators at higher elevations on mountainous terrain, along with harsh weather conditions, has led to the evolution of pollination strategies that are specific to mountain areas.**

Many alpine plants rely on the wind for pollination, producing lightweight and abundant pollen that can be carried over long distances. Others have developed relationships with specialised pollinators, such as certain species of solitary bees adapted to the cold temperatures found in mountainous regions.

7. **Mountain goats, otherwise known as Rocky Mountain goats, have evolved specifically to thrive in the challenging terrain of mountainous regions.**
Renowned for their distinctive appearance, with curved horns and a thick white coat, they exhibit remarkable sure-footedness on steep and rocky surfaces. Their concave hooves provide exceptional traction, allowing them to navigate cliffs with ease. Preferring high-altitude habitats, mountain goats are endemic to the mountainous areas of western North America, including the Rocky Mountains. These herbivores are well-suited to their alpine environments, where they graze on a variety of vegetation, from grasses to alpine shrubs. Only during the harsh winter months do the mountain goats descend to lower elevations in search of more accessible forage. These sure-footed adaptations enable them to access isolated foraging areas and evade potential threats in their high-altitude habitats.

8. **The snow leopard thrives in the remote mountainous regions of Central and South Asia and is well adapted to life in high-altitude environments.**

Snow leopards are superbly adapted predators renowned for their solitary and elusive nature. Their thick, spotted fur provides camouflage in snowy landscapes, and their large nasal cavities and lungs allow for efficient oxygen intake in high-altitude environments. Snow leopards are stealthy hunters known for their ability to ambush prey in the rocky outcrops and cliffs of their mountainous habitats. Preferring to ambush prey from elevated vantage points, they primarily target blue sheep and other mountain ungulates. Unfortunately, the snow leopard faces threats such as habitat loss, poaching, and retaliatory killings, making it an endangered species. In order to protect this unique mountain cat and the delicate environments it lives in, conservation efforts are essential.

9. **Pikas are small, mountain-dwelling, herbivorous mammals related to rabbits and inhabit mountainous regions across Asia and North America.**
These herbivores have round bodies, short limbs, an even coat of fur, and distinctive rounded ears. Pikas play a vital role in their ecosystems by foraging on grasses and other alpine vegetation, creating hay piles by drying and storing vegetation that serves as crucial winter food reserves. Pikas are seen as an indicator species for climate change, as their sensitivity to temperature variations and dependence on cold, rocky habitats make them vulnerable to rising temperatures in mountain ecosystems. Climate change-related rising temperatures

can lead to increased heat stress and reduced foraging time, leading to reduced food reserves.

10. **Golden eagles are birds of prey that inhabit mountainous regions worldwide.**
These birds of prey are adapted to life at high altitudes, with keen eyesight to spot potential prey from great distances. Golden eagles are known for their powerful flight capabilities, allowing them to soar over expansive mountainous landscapes while hunting for small mammals and birds. Their adaptability to various altitudes and diverse prey sources highlights their significance as apex predators in mountain ecosystems.

11. **One of the world's largest flying birds, the Andean condor, is known to soar effortlessly at high altitudes in the Andes Mountains. It uses the heat thermals it passes over to stay in the air, allowing it to soar for hours without beating its wings.**
The Andean condor possesses impressive physical attributes, with a wingspan reaching up to 3.3 metres (10.8 feet) and a weight of up to 15 kilograms (33 pounds). Adapted to its high-altitude habitat, this scavenger bird is known for its remarkable soaring abilities, effortlessly riding thermals and updrafts to soar across vast distances in search of carrion. Despite its large size, the Andean condor is a graceful flyer, capable of reaching altitudes of over 5,000 metres (16,400 feet). Revered by indigenous cultures as a symbol of power and freedom, the Andean condor plays a significant role in Andean mythology and folklore.

Conservation measures are crucial to preserve this iconic species and its mountainous habitat. Unfortunately, habitat loss, poaching, and collisions with power lines pose risks to this magnificent bird.

12. **Mountains significantly influence local and regional weather patterns due to their elevation, topography, and interaction with atmospheric circulation**

    As air masses encounter mountain barriers, they are forced to rise, leading to the phenomenon known as orographic lift - this upward movement of air results in adiabatic cooling of the air, leading to condensation and the formation of clouds. Consequently, mountains often experience higher precipitation levels, particularly on their windward slopes, where moist air is forced to rise. The leeward side, or "rain shadow," typically experiences drier conditions as descending air warms, inhibiting cloud formation and precipitation. Additionally, mountains can create microclimates, with temperature and precipitation variations occurring at different elevations. Higher elevations tend to be cooler and experience more precipitation, while lower elevations may be warmer and drier. These complex interactions between mountains and weather systems significantly affect agriculture, water resources, and ecosystem dynamics in mountainous regions.

13. **Watermelon snow, also known as "pink snow," is a curious phenomenon that occurs on mountains, where patches of snow appear pink or red. It is caused by a cold-loving, green**

microalga called *Chlamydomonas nivalis*, which produces a red pigment to shield itself from the sun's intense UV rays.

The red pigment not only protects the algae but also absorbs heat, causing the surrounding snow to melt just enough to provide the algae with liquid water, which it needs to survive. Interestingly, watermelon snow alters the snow's reflectiveness, causing it to absorb more sunlight rather than reflecting it, and potentially increasing the rate of melt in these areas. Despite watermelon snow being reportedly sweet-smelling, it is not safe to eat, as the algae can cause stomach discomfort.

Mountains and humans

1. **English mountaineer Edward Whymper led a crew that successfully ascended the Matterhorn for the first time in 1865, but four team members perished in the descent.**

   On July 14, 1865, the first climb of the Matterhorn was accomplished successfully and is a historic mountaineering achievement marked by triumph and tragedy. Led by British climber Edward Whymper, the climbing group included seven members, including Swiss guides and British and French climbers. The team ascended the Hörnli Route, which is considered the easiest of the four routes. However, they still faced treacherous conditions and technical challenges. They successfully reached the summit, marking the first time in history that humans had conquered the iconic peak.

However, the descent proved tragic when four members, including experienced Swiss guide Peter Taugwalder and his son, fell to their deaths during a rope-breaking incident. The climb and subsequent tragedy on the Matterhorn captured international attention, highlighting the risks and challenges of alpine mountaineering.

2. **Hannibal's legendary crossing of the Alps in 218 BCE stands as one of the most audacious military feats in ancient history, showing the strategic use of geographical barriers to gain a tactical advantage in warfare.**
The Carthaginian general Hannibal Barca led his army, including war elephants, on a treacherous journey through the rugged and often inhospitable Alps terrain during the Second Punic War against Rome. The exact route remains uncertain, but it is believed that Hannibal navigated challenging mountain passes, enduring harsh weather conditions and facing logistical challenges. Although Hannibal suffered losses during the perilous journey, the successful Alps crossing allowed him to catch the Roman forces off guard, as they were confident in the Alps as a secure natural barrier to invaders, and achieve significant victories at the battles of Trebia, Lake Trasimene, and Cannae.

3. **Oxygen levels decrease with altitude, making it challenging for both humans and animals to adapt to higher elevations.**
Because of the decreased air pressure, high-altitude environments with low oxygen levels pose significant

challenges for humans and animals. As elevation increases, atmospheric pressure decreases, leading to lower oxygen availability. This condition, known as hypoxia, affects the respiratory and cardiovascular systems. Animals in mountainous regions often exhibit specialised adaptations, such as increased lung capacity or more efficient oxygen-carrying capabilities in their blood. Humans, however, face difficulties in adapting quickly to high altitudes. Altitude sickness, with symptoms like headaches, nausea, confusion, and fatigue, can occur as the body is exposed to the reduced oxygen. Prolonged exposure to high elevations may lead to chronic conditions like pulmonary edema or cerebral edema, where fluid excessively builds up in the lungs or brain. Over time, populations living in high-altitude regions, like the Andean or Himalayan communities, have developed genetic adaptations to cope with lower oxygen levels.

4. **Located in the Andes mountains, La Rinconada in Peru is the highest permanent settlement in the world at a height of 5,100 metres (16,700 feet) above sea level.**
Originally established as a gold mining camp, La Rinconada has grown into a makeshift town with a population that fluctuates around 30,000 people. The harsh living conditions include extreme cold, limited oxygen levels, and challenging terrain. Residents endure the difficulties of high-altitude living and engage in artisanal gold mining as a primary source of livelihood.

5. **El Alto is a rapidly growing city adjacent to La Paz, Bolivia. It is one of the highest cities, sitting at around 4,150 metres (13,615 feet) above sea level.**

   In contrast to the small settlement of La Rinconada, El Alto has a highly diverse economy and a population of almost 1 million. Initially established as a peri-urban settlement, El Alto has witnessed significant expansion and urbanisation, evolving into a bustling metropolis. Known for its vibrant street markets, indigenous culture, and unique architectural landscape, El Alto is a major economic hub focusing on trade, commerce, and small-scale industries. The city has a predominantly indigenous population, and its socio-economic development has played a crucial role in shaping Bolivia's contemporary political and cultural landscape.

6. **The "death zone" refers to a high-altitude point on a mountain that is usually above 8,000 metres (26,000 feet) and doesn't have enough oxygen to sustain human life. Both Mount Everest and K2 have summits in the death zone.**

   In a death zone, atmospheric pressure is so low that there is an insufficient amount of oxygen to sustain human life for an extended period; the body uses up its store of oxygen faster than it can be replenished. In this zone, climbers face severe hypoxia, leading to altitude sickness, impaired judgement, and compromised physical abilities. The lack of oxygen increases the danger of life-threatening conditions such as high-altitude cerebral edema (swelling of the brain) and high-altitude pulmonary edema (fluid build-up in the lungs).

Mountaineers attempting to summit peaks in the death zone, such as Mount Everest, must rely on supplemental oxygen to survive, and even with this assistance, prolonged exposure increases the risk of severe health issues.

7. **At high altitudes, the human body undergoes a physiological process called altitude acclimatisation, during which individuals gradually adjust to the reduced oxygen levels.**

   These adaptations include increased breathing rate, more red blood cells produced to enhance oxygen-carrying capacity, and adjustments in blood flow to vital organs. Non-essential bodily functions such as digestion are suppressed. This helps mitigate the risk of altitude sickness, a condition caused by the rapid ascent to elevations above 2,500 metres (8,200 feet). The acclimatisation process typically takes several days to weeks, allowing the body to gradually adjust and reduce the potential negative effects of hypoxia. With acclimatisation and lots of training, it has been possible for some humans to reach the top of Mount Everest without using supplemental oxygen.

8. **It is estimated that 81.6 million people permanently live at heights higher than 2,500 metres (8,200 ft).**

   Over time, populations living at high altitudes, such as those in the Andes or the Tibetan Plateau, have developed genetic adaptations to cope with the challenges of low oxygen. One notable adaptation is the increased production of red blood cells, which enhances the blood's oxygen-carrying capacity. Additionally, efficient respiratory and cardiovascular systems have evolved to better transport available oxygen around the body. These adaptations are believed to be inherited traits that offer advantages in surviving and thriving in oxygen-deprived environments. The adaptation process is a result of natural selection acting over generations, allowing these populations to avoid severe altitude-related illnesses. While individuals not native to such high-altitude regions may experience temporary acclimatisation, the genetic adaptations observed in these populations highlight the remarkable ability of the human body to adjust to challenging environmental conditions over extended periods.

9. **The 1996 Mount Everest disaster stands as one of the most tragic events in the history of high-altitude mountaineering, with 12 people killed due to a blizzard.**
During the climbing season that year, multiple teams attempted to summit Everest, resulting in a series of unfortunate events. On May 10-11, a severe storm struck the mountain, trapping climbers near the summit. Eight climbers, including experienced guides and clients, lost their lives during the descent, and several others suffered severe frostbite and injuries. The disaster gained widespread attention due to the media coverage and the subsequent accounts of survivors, including those documented in Jon Krakauer's book "Into Thin Air." The incident prompted discussions about the commercialisation of Everest, overcrowding, and the ethical considerations of guiding inexperienced climbers

to the world's highest peak. The 1996 Everest disaster remains a sad reminder of the unpredictable and dangerous nature of high-altitude mountaineering.

10. **In 2015, a massive earthquake that struck Nepal triggered an avalanche on Mount Everest that swept through Everest Base Camp, claiming the lives of 22 people and injuring numerous climbers and trekkers.**

    Avalanches, whether natural or human-triggered, are fast-moving masses of snow, ice, and debris that can reach tremendous speeds, causing destruction in their paths. Avalanches are typically influenced by factors such as snowpack conditions, slope angle, and weather patterns. Weak layers in the snowpack can give way under added stress, leading to the release of snow. Avalanche forecasting and mitigation efforts, including controlled releases in certain areas, are essential for managing the risks associated with these natural phenomena. The avalanche struck during the peak of the climbing season and showed the region's vulnerability to natural disasters.

11. **Mountains have held spiritual and religious significance for various cultures throughout history.**

    Many religions attribute sacred qualities to mountains, often regarding them as places where divine beings reside or where important religious events transpired. In Hinduism, Mount Kailash in the Himalayas is believed to be Lord Shiva's abode. Mount Kailash is also considered sacred in Buddhism, Jainism, and the

Tibetan religion of Bon. Mount Sinai is revered in Judaism, Christianity, and Islam, where Moses received the Ten Commandments. According to Native American mythology, some mountains are sacred and connected to creation tales and ceremonies. It was also thought that the Greek gods lived atop Mount Olympus.

12. **The Yeti, also known as the "Abominable Snowman," is a mythical ape-like creature that is supposed to live in the remote, mountainous regions of the Himalayas.**

    Described as large, hairy, and humanoid, the Yeti has become a legendary figure in Nepal, Tibet, and Bhutan folklore. While there is no scientific evidence supporting the existence of the Yeti, numerous anecdotal accounts, footprints, and alleged sightings have fueled widespread interest and speculation. Various cultures in the Himalayan region have their own names and beliefs regarding this elusive creature. Many expeditions and searches for the Yeti have been conducted over the years, often fueled by both curiosity and the desire to prove or debunk its existence. The Yeti remains an enduring part of Himalayan mythology, blending mystery, adventure, and cultural fascination.

13. **Mountains serve as water towers and sources for numerous rivers and streams, supplying half of the world's population with water.**

    When snow and ice from mountainous regions melt, it creates essential water sources that support ecosystems and communities downstream. Many of the world's major rivers, such as the Ganges, Indus, Yangtze, and

Colorado, originate in mountainous areas. Millions of people who live in lowland areas depend on these rivers for their supply of fresh water for industrial, drinking, and agricultural uses. However, the delicate balance of mountain ecosystems is vulnerable to climate change, leading to alterations in precipitation patterns and an acceleration of glacial melt.

14. **One of the most harrowing survival stories in aviation history is that of Uruguayan Air Force Flight 571, which crashed in the Andes Mountains in 1972.**

    Flying planes over mountains presents a unique set of challenges and potential dangers for pilots. Unpredictable weather conditions, such as high winds, turbulence, and rapidly changing visibility, pose a severe risk. The Uruguayan Air Force Flight tragedy unfolded when the aircraft encountered severe weather conditions, with poor visibility and high winds. After a failed attempt to navigate through the mountains, the aircraft ultimately collided with the mountainside at an altitude of over 11,800 feet. Miraculously, despite the devastating impact, 27 passengers initially survived the crash. However, stranded in the harsh, frozen landscape with limited provisions, the survivors faced unimaginable challenges, including extreme cold, avalanches, and the absence of rescue. Resorting to drastic measures for survival, including cannibalism, the survivors endured a gruelling 72 days before a small group of them was finally rescued. In response to this incident and others like it, aviation authorities continuously emphasise the importance of pilot training, situational awareness, and the use of advanced technology to reduce the dangers associated with flying over mountains.

15. **Due to gravitational time dilation, time passes at a slightly faster rate at higher altitudes than it does at sea level. This effect, predicted by Einstein's theory of relativity, has been confirmed by precise measurements using atomic clocks placed at different elevations.**

According to general relativity, gravitational fields affect the passage of time, with time passing more slowly at lower altitudes where gravity is stronger due to it being closer to the Earth's core. Since Earth's gravitational field is weaker at higher altitudes, such as on top of mountains, clocks at higher elevations experience less gravitational time dilation than clocks at lower elevations. This means that time actually moves slightly faster at higher altitudes than at sea level. While the effect is minuscule and imperceptible in everyday life, it has been confirmed through precise measurements using atomic clocks placed at different elevations. Therefore, individuals living or working in mountainous regions may experience time dilation effects, although they are typically negligible in practical terms. Nonetheless, the phenomenon offers fascinating insights into the interplay between gravity, spacetime, and the perception of time in different environments.

16. **Low-light pollution and reduced atmospheric interference make high-altitude observatories**

atop isolated mountain peaks ideal for astronomical observations.
These elevated locations offer optimal conditions for astronomical observations due to several factors:

1. The thin, dry atmosphere at high altitudes reduces atmospheric interference, such as light pollution and atmospheric distortion, making images of celestial objects clearer and more distinct.
2. Mountaintop observatories benefit from stable atmospheric conditions, which reduce the effects of atmospheric turbulence on telescope performance.
3. The high elevation provides astronomers with access to a greater portion of the atmosphere, allowing observations across a broader range of wavelengths, including infrared and submillimetre radiation.

Leading mountaintop observatories around the world, like the Atacama Large Millimetre Array (ALMA) in Chile and the Mauna Kea Observatory in Hawaii, are conducting ground-breaking research on a variety of subjects, from exoplanets and stellar evolution to cosmology and galaxy formation. Their findings are essential to expanding our understanding of the universe.

17. **Jordan Romero, who reached the summit at the age of 13 in 2010, is the youngest person to have climbed Mount Everest; the oldest person to**

have climbed Everest is Yuichiro Miura of Japan, who achieved the feat at 80 in 2013.

Jordan was accompanied by his father, stepmother, and a team of experienced guides during the expedition. His successful summit of Everest was the culmination of an incredible journey that began with his passion for climbing at a young age. Since then, Jordan has continued his mountaineering adventures, completing the Seven Summits (the highest peak on each continent) at the age of 15.

When renowned Japanese mountaineer, skier, and adventurer Yuichiro Miura set the record for the oldest person to summit Everest, it was not his first ascent. He had previously summited the mountain twice, in 2003 and 2008. Yuichiro Miura's successful summit at the age of 80 inspired people worldwide, demonstrating that age is no barrier to achieving extraordinary feats of endurance and determination. His accomplishment highlighted the importance of physical fitness, mental resilience, and careful preparation in tackling the challenges of high-altitude mountaineering.

18. **The first flight over Mount Everest was a historic aviation achievement that took place on April 3, 1933.**

The flight was part of the Houston Mount Everest Flight Expedition, which aimed to conduct aerial surveys of the Himalayan region and gather scientific data, including photographs and measurements of the world's highest peak. The flight was piloted by British Royal Air Force (RAF) Flight Lieutenant David McIntyre, with

American aviator Colonel Lawrence W. Houston as a passenger. Flying a Westland PV-3 biplane named "Houston-Westland," despite encountering challenges, the expedition successfully flew over the summit of Mount Everest at an altitude of approximately 9,500 metres (31,200 feet), setting a new altitude record for powered flight at the time. This first flight over Mount Everest provided valuable insights into the geography, topography, and weather patterns of the Himalayan region. The expedition's aerial photographs and scientific data contributed to our understanding of Earth's highest peaks and their surrounding landscapes. While the flight lasted only a few hours, its impact on aviation and exploration was profound, inspiring future generations of pilots, adventurers, and scientists.

19. **Mountains offer potential opportunities for renewable energy generation, including hydropower from mountain rivers and streams, geothermal energy from volcanic activity, and wind energy from high-altitude wind currents. These sources of energy can contribute to sustainable development in mountain regions.**

One of the primary sources of energy in mountainous areas is hydropower, derived from the gravitational potential energy of water stored in mountain rivers and reservoirs. Hydroelectric dams utilise the kinetic energy of flowing water to drive turbines and generate electricity, providing a reliable and environmentally friendly power source. Mountainous areas also have potential for other renewable energy sources, such as solar and wind energy. Wind turbines can be strategically located on mountain ridges and passes to capture the strong, consistent winds prevalent at higher elevations.

Similarly, solar panels installed on mountain slopes and rooftops can harness the abundant sunlight to generate electricity. Geothermal energy, produced from heat stored beneath the Earth's surface, is another potential energy source in mountainous areas with volcanic activity or geothermal hotspots. Mountain energy harvesting reduces reliance on fossil fuels and promotes sustainable development, job creation, and energy independence in remote mountain communities. However, in order to minimise the impact on fragile mountain ecosystems and maintain the biodiversity and natural beauty of mountain landscapes, rigorous planning and environmental considerations are essential.

20. **Mountainous terrain provides the backdrop for gruelling marathon races, such as the Ultra-Trail du Mont-Blanc (UTMB) in the Alps and the Hardrock 100 in the Colorado Rockies. These races test runners' endurance and mental fortitude across challenging trails.**

    Unlike traditional marathons, which are 42 kilometres (26 miles) long, ultra-marathons typically exceed this distance, ranging from 50 kilometres (31 miles) to 100 kilometres (62 miles) or more. The terrain encountered in mountain ultra marathons is often challenging, with steep ascents and descents, rocky trails, river crossings, and unpredictable weather conditions. These races demand exceptional physical fitness, mental fortitude, and strategic planning from participants. On average, it can take elite runners anywhere from 8 to 24 hours to complete a mountain ultra-marathon, depending on factors such as course difficulty, elevation gain, and weather conditions. Some runners take more than 24 hours to complete the ultra-marathons, meaning they must run through the night. Participants face altitude sickness, dehydration, muscle fatigue, blisters, and navigation errors. Despite these obstacles, mountain ultra marathons attract a dedicated community of trail runners and outdoor enthusiasts drawn to the adventure, camaraderie, and personal achievement that comes from pushing the boundaries of human endurance amidst breathtaking mountain landscapes.

21. **Mount Vesuvius, located in Italy near Naples, is one of the most dangerous volcanoes as it is still active and has a population of 3 million people living near its base.**
Best known for its catastrophic eruption in 79 AD, Vesuvius buried the Roman cities of Pompeii and Herculaneum under thick layers of ash and pumice, preserving them for centuries. This eruption released a deadly cloud of gas, molten rock, and ash high into the atmosphere, killing thousands of people. Vesuvius is a stratovolcano, meaning it has steep, conical slopes built from layers of hardened lava, ash, and volcanic debris. The volcano has erupted more than 30 times since 79 AD, with its most recent eruption in 1944. Scientists closely monitor Vesuvius for signs of future activity, and emergency plans have been put in place, as its eruptions have the potential to be extremely destructive.

Emergency plans involve the evacuation of the 700,000 people living in the "red zone", an area where the risk of pyroclastic flows is highest.

Forests

Types of forest

1. **The United Nations Food and Agriculture Organization (FAO) defines a forest as "land spanning more than 0.5 hectares with trees higher than 5 metres and a canopy cover of more than 10 percent, or trees able to reach these thresholds in situ".**
A dense collection of trees, plants, fungi, and a rich variety of wildlife are what define a forest. They serve as vital habitats for a wide variety of plant and animal species, support global biodiversity, and contribute significantly to climate regulation by absorbing carbon dioxide and releasing oxygen. Forests are not only essential for environmental health but also hold cultural, economic, and recreational significance for communities around the world.

2. **Approximately 31% of the surface area of Earth's land is covered by forests.**
This is roughly 4 billion hectares (10 billion acres). These expansive wooded regions are distributed across

various climates and continents, ranging from tropical rainforests near the equator to boreal forests in the northern latitudes.

3. **By absorbing and storing large volumes of carbon dioxide, forests function as carbon sinks, reducing the effects of climate change.**
Through photosynthesis, trees absorb CO_2 and release oxygen, contributing to the planet's oxygen supply. The stored carbon in forests helps regulate the Earth's climate by mitigating the impacts of climate change.

4. **Tropical rainforests, situated near the equator, are among the Earth's most biodiverse and vital ecosystems. They are home to approximately 50% of the world's known biodiversity.**
These lush environments, characterised by high temperatures and abundant rainfall, contain an astonishing array of plant and animal species. Numerous species live in their complex, multi-layered environment, which is created by the dense canopies of their tall trees. In addition to their ecological significance, tropical rainforests benefit indigenous populations and the health of the planet.

5. **The Amazon Rainforest, which spans nine South American countries, with the majority in Brazil, Peru, and Colombia, is the world's biggest tropical rainforest.**
Covering approximately 6.7 million square kilometres, the Amazon Rainforest is home to many plants and animals not found anywhere else on the planet. The

Amazon is crucial in regulating the global climate by producing about 20% of the world's oxygen and absorbing a significant amount of carbon dioxide.

6. **Boreal forests, also known as taiga, form one of the largest terrestrial biomes, extending across vast regions of Canada, Alaska, Scandinavia, and Russia.**
Characterised by cold temperatures and dominated by coniferous trees such as spruce, pine, and fir, the forest is another type of forest that is crucial in the global carbon cycle. Its expansive tree cover helps hold substantial amounts of carbon dioxide. The boreal forest's lakes, wetlands, and rivers offer vital habitats for fish and waterfowl, while the forests contain iconic animals like moose, wolves, and bears.

7. **Surprisingly, diseases, insects, and forest fires all contribute to the regeneration and health of boreal forests. Despite their apparent destructiveness, these disturbances are crucial in preserving the forest ecosystem by recycling nutrients and supporting natural regeneration.**
Fires consume old vegetation, releasing nutrients like nitrogen, phosphorus, and potassium back into the soil, creating fertile ground for new plants, and opening up space for young trees to grow. In fact, some boreal trees, like jack pine, have cones that only release seeds under the intense heat of a fire, making flames essential to their life cycle. Diseases and insect infestations, on the other hand, target older or weaker trees, naturally thinning the forest and reducing competition for light,

water, and nutrients. Insects like bark beetles even help recycle nutrients by breaking down wood, while dead and decaying trees from disease become new habitats for birds, small mammals, and countless decomposers.

8. **Not only found in tropical regions, rainforests can also exist in temperate zones, such as the Pacific Northwest's temperate rainforests.**
Temperate rainforests, found between the tropical and boreal regions, are lush ecosystems characterised by high levels of rainfall, mild temperatures, and dense vegetation. Unlike tropical rainforests, temperate rainforests are found in cooler climates, typically along the western coasts of continents. Notable examples include the Pacific Northwest in North America and parts of Chile, New Zealand, and Australia. Towering conifers like Sitka spruce, Douglas fir, and redwoods create an incredible canopy in these habitats. The undergrowth is rich with many different plant species, including ferns and mosses. However, these ecosystems face threats from logging and habitat fragmentation.

9. **Mangrove forests, found along the coast where freshwater and saltwater coexist, are home to 80 different varieties of mangrove trees.**
As mangrove forests are situated in intertidal zones, which are the areas between the high and low tide marks, they display remarkable adaptability for high salinity, low environmental oxygen levels, and frequent tidal flooding. These salt-tolerant trees and shrubs form dense thickets along tropical and subtropical coastlines. Mangroves function as natural buffers, protecting

shorelines from erosion and acting as nurseries for a variety of marine species. Their intricate root systems stabilise coastal sediments and create habitats for diverse aquatic life.

10. **The Sundarbans, located in the delta region of the Padma, Meghna, and Brahmaputra river basins in Bangladesh and India, constitute the largest mangrove forest in the world, spanning approximately 10,000 square kilometres.**
The Sundarbans are a complex network of tidal waterways, mudflats, and small islands characterised by their extensive mangrove tree cover. They are renowned for their rich biodiversity, housing many different animals, including crocodiles, Bengal tigers, spotted deer, as well as numerous bird species. The mangrove ecosystem provides a crucial buffer against storm surges and supports the livelihoods of local communities through fishing and honey collection.

11. **Eucalyptus forests, primarily found in Australia, are dominated by trees belonging to the Eucalyptus genus, which comprises over 700 species.**
Eucalyptus leaves contain oil glands which release aromatic eucalyptus oil. Characterised by tall, slender trees with unique bark patterns, these forests are resilient and have adapted to fire-prone environments, some species being capable of resprouting after a fire. The leaves contain volatile oils that contribute to the characteristic blue haze often seen around these forests. The tree's volatile oils can ignite, promoting controlled

burns that clear the forest floor of accumulated debris and stimulate germination. These forests also support diverse wildlife, including koalas that exclusively feed on eucalyptus leaves. Eucalyptus forests contribute to Australia's timber industry, providing valuable hardwood for construction and paper production. The rapid growth of eucalyptus trees makes their ecological significance even more valuable.

12. **Cloud forests are unique and enchanting ecosystems known for their persistent low-level cloud cover, creating an otherworldly atmosphere.**
Typically found in mountainous regions, these forests are often situated at elevations where clouds envelop the landscape, providing a constant source of moisture. The combination of high humidity, cool temperatures, and abundant cloud water promotes the growth of mosses, ferns, and epiphytic plants, creating a lush and diverse environment. Cloud forests are vital for maintaining water balance in their respective regions, capturing moisture from passing clouds and releasing it gradually, thus preventing rapid runoff and erosion. These ecosystems harbour an extraordinary number of animal and plant species, many of which are endemic and adapted to the unique conditions. A notable example is a species of bird called the resplendent quetzal found in the Central American cloud forests.

13. **The Black Forest is known for its dense canopy, which can be so thick that it earned its name**

due to the reduced sunlight penetrating the forest floor.

The Black Forest, or Schwarzwald in German, is a captivating region in southwestern Germany renowned for its dense, dark canopy of evergreen trees, primarily spruce and fir. Covering an area of about 6,000 square miles, the forest is famous for its picturesque landscapes and charming villages. In addition to its inherent charm, the Black Forest is known for its delicious products, such as Black Forest ham, Black Forest cake, and traditional cuckoo clocks. The forest contains hiking trails, waterfalls, and opportunities to experience the region's cultural heritage in charming towns like Triberg and Freiburg.

14. **Well-known for its connection with the legend of Robin Hood, Sherwood Forest is home to the Major Oak. This massive oak tree is said to have sheltered Robin Hood and his Merry Men.**

    Sherwood Forest, located in Nottinghamshire, England, is an iconic woodland with a rich history and legendary associations. Covering approximately 450 acres, Sherwood Forest is famous for being the alleged home of the legendary outlaw Robin Hood and his band of Merry Men. The forest's most notable inhabitant is the Major Oak, a massive oak tree estimated to be around 800 to 1,000 years old, which, according to folklore, served as Robin Hood's shelter. Sherwood Forest has been a royal hunting ground since the medieval period and is now a designated country park.

15. **Kelp forests are a type of forest that is found underwater and are dominated by large, brown algae known as kelp.**
These underwater forests thrive along temperate coastlines worldwide, playing a crucial ecological role. Giant kelp, a species often found in these forests, can grow up to two feet per day, reaching heights of over 100 feet. Kelp provides essential habitat and serves as a sanctuary for lots of marine life, including fish, invertebrates, and sea otters. The intricate structure of kelp forests offers shelter and protection, and the kelp itself supports a complex food web.

16. **Tongass National Forest, located in southeastern Alaska, is the largest national forest in the United States and the world's largest temperate rainforest.**
Encompassing approximately 16.7 million acres, the forest consists of towering mountains, pristine fjords, and lush rainforests. Home to diverse ecosystems, including old-growth stands of Sitka spruce and Western hemlock, the Tongass supports an abundance of wildlife, including iconic species such as bald eagles, brown bears, and Pacific salmon. The forest has significant cultural importance to indigenous communities, with numerous archaeological sites reflecting the rich history of the Tlingit, Haida, and Tsimshian peoples.

17. **Białowieża, situated on the border between Poland and Belarus, is home to one of the last**

remaining intact areas of the ancient forest that once covered much of Europe.

This UNESCO World Heritage site spans approximately 580 square miles and is renowned for its ancient and untouched character, dating back to the last Ice Age. The forest is home to the European bison, the continent's heaviest land animal, which was successfully reintroduced into the wild here. The towering trees, including massive oaks and spruces, create a unique habitat supporting a rich diversity of flora and fauna. Białowieża Forest has deep cultural and historical roots, having served as a hunting ground for Polish kings and Russian tsars. Today, it is a crucial centre for scientific research and nature conservation, attracting visitors with its pristine wilderness and the sense of stepping back in time to a truly ancient European landscape.

18. **The Daintree Rainforest, located in Queensland, Australia, is one of the oldest rainforests in the world. It is estimated to be around 180 million years old.**

    Spanning approximately 1,200 square kilometres, it is part of the Wet Tropics of Queensland, which is found along the northeast coast of Australia. The Daintree is renowned for its remarkable biodiversity, supporting many plant and animal species, many of which are unique to this area and are found nowhere else on the planet. The forest is home to ancient plant lineages, primitive flowering plants, and a diverse range of wildlife, including the endangered cassowary and the vibrant Boyd's forest dragon. The Daintree Rainforest also features crystal-clear streams, secluded beaches, and unique ecosystems like the mangrove swamps along the coastline. In addition to its ecological significance, the Daintree holds cultural importance for the local Kuku Yalanji Aboriginal people, who have a deep connection to the land.

19. **Yakushima Forest, located on Yakushima Island in Japan, is renowned for its ancient cedar trees, some of which are over 1,000 years old.**
The most famous among them is Jomon Sugi, an ancient cedar estimated to be between 2,170 and 7,200 years old. The forest's moss-covered rocks, lush vegetation, and ancient trees create a magical and serene atmosphere. The unique ecosystem of Yakushima includes numerous plant and animal species, and the forest inspired Hayao Miyazaki's animated film, Princess Mononoke. The island's remoteness has contributed to the preservation of its pristine environment, and Yakushima Forest continues to captivate visitors with its ancient beauty.

20. **Located in British Columbia, Canada, Great Bear Island is the biggest temperate rainforest in the world, with a total area of approximately 6.4 million hectares.**
This ecologically rich and biodiverse region is characterised by towering ancient trees, pristine fjords, and a complex network of rivers and estuaries. The forest is home to many wildlife, including the rare white Kermode bear, also known as the spirit bear, which holds cultural significance for the area's indigenous peoples. Grizzly bears, wolves, salmon, and numerous

bird species contribute to the remarkable biodiversity of the Great Bear Rainforest. The region has been a focal point for conservation efforts, balancing sustainable resource management with preserving its unique ecosystems. Indigenous communities, who have been caring for these lands for generations, are essential in safeguarding the Great Bear Rainforest, which has gained international recognition as an example of sustainable environmental practices.

21. **Part of Olympic National Park, the Hoh Rainforest is one of the wettest places in the continental United States. It receives about 3.6 metres (over 12 feet) of rain annually, contributing to its lush and vibrant ecosystem.**

    The Hoh Rainforest, situated in Olympic National Park on the Olympic Peninsula in Washington, USA, is a captivating example of a temperate rainforest. This lush and verdant ecosystem receives an annual rainfall of up to 12 feet, making it one of the wettest places in the continental United States. The Hoh Rainforest contains towering old-growth trees, including Sitka spruce and Western hemlock, draped in mosses and ferns. The Hall of Mosses Trail offers visitors an immersive experience, winding through the ethereal landscape of trees and hanging mosses. The forest is home to many animal species, such as Roosevelt elk, black bears, and various bird species. Beyond its ecological significance, the Hoh Rainforest is a living laboratory for scientists studying temperate rainforest ecosystems.

22. **Kakamega Forest, located in western Kenya, is the last remaining tropical rainforest in the country.**

   This ancient forest holds a remarkable biodiversity, covering an area of about 238 square miles. It is often referred to as the "last remnant of the Guineo-Congolian rainforest that once spanned the continent." Kakamega Forest contains towering trees, lush vegetation, and a rich variety of flora and fauna. It provides an essential home for a wide range of wildlife, such as primates like colobus monkeys and the endangered De Brazza's monkey.

23. **The Chemnitz Petrified Forest, located near the city of Chemnitz in Saxony, Germany, is an extraordinary geological site that contains petrified trunks from trees and other plants.**

   Dating back to the Carboniferous period, approximately 291 million years ago, this petrified forest preserves fossilised remains of trees and plants, turning them into stone over thousands of years. The forest was formed due to the eruption of a nearby volcano where trees and plants were covered by hot tephra (volcanic rock and lava). The volcanic ash contained silica-rich minerals, which seeped into the buried wood and replaced the organic matter over millions of years. This process, called petrification, preserved the intricate details of the plant cells, turning wood into stone. Over time, geological forces like erosion gradually exposed the petrified wood. The first scientific record of the Chemnitz fossils dates back to 1546, and excavations have continued ever since, revealing a large number of fossils. Notably, the Chemnitz Petrified Forest is known for its well-preserved specimens of tree ferns, calamites (a type of ancient plant related to horsetails), and other extinct plant species. The site's museum showcases a remarkable collection of petrified wood, allowing visitors to view the intricate details of these ancient plants.

Ecosystems

1. **The Amazon Rainforest is home to an estimated 390 billion individual trees representing around 16,000 different species and is home to approximately 10% of all known species on Earth.**
   This remarkable diversity of life within the Amazon Rainforest is not limited to just trees; it also includes other plants and animals. The forest floor is home to a wide range of plant species, while the canopy supports an incredible variety of animal life, including over 1,300 bird species, 427 mammals, and a staggering 2.5 million insect species. Jaguars, harpy eagles, and Amazon river dolphins are just a few of the species that are part of the complex web of life in the Amazon.

2. **Indigenous people in the Amazon have long utilised the diverse plant life for medicinal purposes. Around 25% of modern pharmaceuticals are estimated to originate in rainforest plants.**

Quinine is a medication found in the bark of the Cinchona tree that is crucial for treating malaria. Additionally, the rosy periwinkle plant produces compounds essential in treating certain cancers. Plants like the Cat's Claw vine are known for their anti-inflammatory properties, while the Copaiba tree produces an oil with anti-inflammatory and antibacterial qualities. These are just a few examples; further studies are being conducted to uncover new medicinal wonders within the Amazon Rainforest.

3. **The tallest tree in the world is called Hyperion, a coast redwood, located in a remote area of Redwood National Park in California, USA. It reaches an astonishing height of approximately 115.7 metres (379.7 feet).**
Hyperion was discovered in 2006 by naturalists Chris Atkins and Michael Taylor. It holds the record for the tallest living tree on the planet and is estimated to be between 600 and 800 years old. The exact location of Hyperion is kept confidential to protect the tree from human impact. However, despite these efforts, some determined individuals have managed to locate Hyperion. As there is no clear hiking route to the tree, those determined to reach it have had to cut their way through dense vegetation, which has also led to visible signs of soil erosion and damage to the forest. Anyone now attempting to reach Hyperion faces a fine of $5,000 and potential jail time.

4. **The overlapping branches and leaves of towering trees create the rainforest canopy, a**

rich and complex ecosystem located high above the forest floor. **The canopy layer of the rainforest is thought to support between 60 and 90 percent of all life in the rainforest.**

This upper layer of the rainforest, often referred to as the "roof of the rainforest," can reach up to 30 metres (100 feet) or more above the ground and is characterised by a dense network of vegetation and an extraordinary diversity of plant and animal life. The canopy is a crucial habitat for numerous species, including birds, insects, mammals, and epiphytic plants. Adapted to life in the canopy, these organisms have evolved specialised behaviours and physical features to navigate the complex three-dimensional environment. The canopy also plays a vital role in regulating the microclimate of the rainforest, influencing factors such as temperature, humidity, and light availability.

5. **The Amazon Rainforest generates its own water system by releasing water vapour through a process known as transpiration.**

The vast expanse of trees in the rainforest releases an enormous amount of water vapour into the atmosphere through their leaves. This process, along with evaporation from the forest floor, forms massive cloud formations. These clouds contribute to the region's high humidity and frequent rainfall, creating a self-sustaining water cycle unique to the Amazon. This water system not only supports the plants and animals that live within the rainforest but also influences weather patterns on a global scale. This moisture produced by the rainforest also helps sustain rainfall in other parts of

South America through "flying rivers", which occur when winds carry away the moisture.

6. **Victoria amazonica, commonly known as the Amazon water lily or Victoria water lily, is the world's largest water lily that can reach up to 3 metres (10 feet) in diameter.**
The leaves have an upturned rim and a structure that allows them to support the weight of a small child. The Victoria amazonica has a unique method of reproduction: its large white flowers bloom at night and release a sweet scent that attracts pollinators, primarily scarab beetles. As the flowers age, they turn pink before closing. The enormous leaves, with their ability to repel water due to a layer of microscopic bumps, create a natural buoyancy that allows them to stay afloat. The stem and underside of the leaves are covered in many tiny spines to defend themselves from underwater creatures.

7. **Found in forests, carnivorous plants have evolved to trap and digest their prey, typically insects, to increase their nutrient intake.**
These plants have adapted to nutrient-poor environments, such as the Venus flytrap found in boggy areas of the southeastern United States. The Venus flytrap's modified leaves snap shut when triggered by insects, trapping and digesting them with specialised enzymes. Another fascinating carnivorous plant is the pitcher plant, often found in tropical rainforests. These plants lure insects into a tubular pitcher filled with

digestive fluids, extracting essential nutrients to compensate for the nutrient-deficient soils.

8. **Stick insects and leaf-tailed geckos have mastered the art of camouflage by resembling the vegetation in which they inhabit.**
Stick insects, also known as walkingsticks, have elongated bodies and a remarkable resemblance to twigs or branches, allowing them to evade predators by mimicking the vegetation they live in. Similarly, leaf-tailed geckos in tropical rainforests exhibit uncanny leaf-like extensions and colouration, enabling them to merge seamlessly with the foliage. These camouflage tactics serve as both defensive mechanisms and tools for ambushing prey.

9. **The Sumatran orangutan is a critically endangered great ape species native to the rainforests of Sumatra, Indonesia.**
Recognised for its distinctive long, shaggy reddish-brown fur and unique facial features, the Sumatran orangutan is one of the two existing species of orangutans. These arboreal primates are highly adapted to life in the treetops, using their long arms to swing from branch to branch in search of fruits, leaves, and insects. Habitat loss due to illegal logging, agricultural expansion, and human-wildlife conflict severely threaten their survival. To maintain the declining population of Sumatran orangutans in the wild, conservation efforts—such as the creation of protected areas and programmes to combat deforestation—are essential.

10. **Foxfire is a phenomenon that refers to the bioluminescent light emitted by certain species of fungi in forests.**

    Among the most well-known foxfire fungi is the genus Armillaria. These mushrooms produce a soft, eerie glow caused by the chemical reaction between luciferin and oxygen, mediated by the enzyme luciferase. This bioluminescence's purpose is to attract insects, aiding in the dispersal of the fungus's spores. Foxfire is often observed in decaying wood and damp, shaded areas, adding a mystical and enchanting quality to nocturnal forest landscapes. While it poses no harm to humans, the luminescent glow of foxfire fungi continues to captivate both scientists and nature enthusiasts.

11. **Epiphytic plants are species that thrive by growing on the surface of other plants, utilising them as support structures without relying on them for nutrients.**

    Commonly found in tropical rainforests, epiphytic plants, such as orchids, ferns, cacti, and bromeliads, secure themselves to trees or other vegetation. Elevated off the ground, epiphytes obtain water and nutrients from the surrounding air, rain, and organic matter. To improve their ability to absorb moisture and nutrients, they often have adaptations like modified leaves or specialist roots. Epiphytic plants contribute to the rich biodiversity of forest ecosystems, creating microhabitats in the canopy for various organisms.

12. **Some forest animals, such as the oxpecker bird and rhinoceros, have symbiotic relationships. The bird feeds on ticks and parasites on the rhino's skin.**

    The link between the rhinoceros and the oxpecker bird is an example of how symbiotic relationships frequently thrive in forest ecosystems. This unique partnership is defined by mutual benefits: as the oxpecker perches on the rhino's back, it diligently feeds on ticks and parasites that infest the rhino's skin. In return, the rhinoceros gains relief from the parasites that may cause irritation and discomfort. The oxpecker serves as a vigilant exterminator, contributing to the rhino's overall well-being, while the rhino, in turn, provides a convenient and mobile feeding ground for the bird. Furthermore, oxpeckers serve as a natural early-warning system, alerting the rhinos to potential threats or approaching predators. This is especially useful for rhinos as they have very poor eyesight and can rely on oxpecker's warning calls to get themselves out of harm.

13. **Giant pandas are iconic and endangered members of the bear family. They are native to the mountainous bamboo forests of central China.**

    Despite being classified as carnivores, the giant panda's diet is primarily herbivorous, consisting almost entirely of bamboo, supplemented occasionally with fruits and other vegetation. Giant pandas possess a distinctive pseudo-thumb, an extension of the wrist bone, which helps them hold bamboo stems and leaves with ease. Despite their formidable size, giant pandas are generally peaceful animals, spending a significant portion of their day feeding and the remainder resting. Giant pandas face significant conservation challenges due to habitat loss, fragmentation, and human activities. Extensive efforts, including habitat preservation and breeding programs, aim to secure the panda's survival and protect their unique ecosystems.

14. **Arboreal animals, like tree frogs and koalas, spend most of their lives in trees, adapting to life in the canopy of forests.**
This diverse group includes many species, such as monkeys, squirrels, tree frogs, and birds. Their specialised adaptations for arboreal living often include prehensile tails, grasping hands or feet, and keen agility for navigating branches. By using the trees as refuge and platforms for foraging, breeding, and evading predators on the ground, these animals have adapted to take advantage of the three-dimensional space of their elevated homes. The arboreal lifestyle provides numerous advantages, including access to food resources, safety from ground threats, and efficient travel through the interconnected network of branches.

15. **Forests are efficient recyclers. Decomposers break down fallen leaves and dead organisms, enriching the soil with nutrients for new plant growth.**
Microorganisms, including fungi, bacteria, and insects, break down organic matter on the forest floor. These decomposers return essential nutrients to the soil by breaking down complex organic compounds. The

recycled nutrients become available for plant uptake, facilitating new growth and sustaining the overall health of the forest. This natural recycling process maintains the balance of nutrients within the ecosystem and contributes to soil fertility.

16. **The Amazon Rainforest is home to an estimated 2.5 million different insect species.**
Each of these many species—which include beetles, butterflies, ants, mosquitoes, and innumerable others— plays a crucial part in pollination, decomposition, and the cycling of nutrients. These insects perform a variety of tasks that are essential to life in the Amazon, from breaking down organic debris on the forest floor to acting as a food source for many larger species.

17. **The mycorrhizal network is a complex underground network found in forests, formed by mycorrhizal fungi connecting with the roots of plants.**
This intricate relationship facilitates a mutualistic exchange of resources. The mycorrhizal fungi colonise the plant roots, extending their hyphal networks into the soil and enhancing their ability to absorb water and nutrients, particularly phosphorus and nitrogen. In return, the host plants provide the fungi with sugars produced through photosynthesis. Both plants and fungi associate with multiple symbiotic partners at once, and both plants and fungi are capable of preferentially allocating resources to one plant over another.

18. **Forests are critical in preventing soil erosion, a natural process which is made worse by human activities.**

    The intricate root systems of trees and plants in forests act as a natural binding agent, anchoring soil and preventing its displacement by wind or water. The canopy formed by the forest cover helps shield the soil from the impact of rainfall, reducing the force of water droplets and minimising surface runoff. Additionally, the layer of decomposing leaves and organic matter on the forest floor enhances soil structure and water absorption. When forests are cleared or degraded, the loss of this protective cover often leads to increased erosion, soil degradation, and the heightened risk of landslides.

19. **Trees possess a fascinating form of "environmental memory" that enables them to remember and respond to past climatic events, helping them better survive future challenges.**

    This memory is stored at the cellular level, particularly in the tree's growth rings and its physiological processes. For example, when a tree experiences drought, it often narrows its growth rings and adjusts its water-use efficiency to conserve resources. If drought conditions recur, the tree "recalls" its prior adaptation, allowing it to respond more effectively by reducing water loss and shifting energy to essential survival functions. Research into tree memory is helping scientists understand how forests might adapt to climate change, revealing that trees with strong environmental memory could play a vital role in maintaining forest health under increasingly variable conditions.

Forests and humans

1. **Despite its vastness, a significant portion of the Amazon Rainforest remains unexplored, with scientists regularly discovering new species of plants and animals.**
   These unexplored regions are also often inhabited by indigenous tribes that have had minimal contact with the outside world. The unexplored landscapes in these areas hold great scientific potential, offering opportunities for groundbreaking discoveries and insights into the functioning of one of the world's most complex and diverse ecosystems. No one knows the actual amount of unexplored territory, but technology is making it easier for scientists to examine these remote regions. In 2022, remote sensing devices discovered ancient settlements in the Bolivian Amazon.

2. **The Amazon Rainforest is home to around 400 different indigenous tribes, some of which have had minimal contact with the outside world.**
   These tribes often possess unique languages, cultures, and sustainable ways of living. They have coexisted harmoniously with the rainforest for centuries, relying on its abundant resources for both medical knowledge and food. Indigenous tribes play a vital role in maintaining the ecological balance of the Amazon, practising sustainable ways of living that prioritise

environmental conservation. Many tribes face increasing dangers from deforestation, encroachment on their ancestral lands, and exposure to external diseases. Despite these challenges, these communities persist in their efforts to protect their cultural heritage and the invaluable biodiversity of the Amazon Rainforest.

3. **Deforestation, brought on by human activities such as logging, agriculture, and infrastructure development, has destroyed around 17% of the Amazon Rainforest.**
This widespread destruction greatly contributes to global climate change and jeopardises the rainforest's biodiversity. When trees are burned or cut down, the carbon they contain is released back into the atmosphere as CO_2, escalating the greenhouse effect and contributing to global warming. The consequences extend beyond regional concerns, impacting weather patterns, biodiversity, and indigenous communities. Forest restoration and preservation are therefore crucial tactics in the battle against climate change.

4. **Scientists have identified the "Amazon tipping point" as a critical threshold beyond which the rainforest may experience irreversible damage. If deforestation reaches 20-25% of the Amazon's total area, it could trigger irreversible ecological damage.**
Beyond this point, the rainforest may enter a feedback loop, experiencing widespread dieback, increased vulnerability to wildfires, and change into a more arid,

savanna-like ecosystem. This tipping point emphasises how urgently strong conservation efforts are needed to stop deforestation, as crossing this threshold could have serious, far-reaching consequences for the Amazon's unique biodiversity and indigenous communities and the global climate.

5. **After disturbances such as natural disasters or logging, rainforests can regenerate through a process called secondary succession.**
Pioneer species are the most resilient, fast-growing plants and trees that are the first to grow after an area has been destroyed by an event such as wildfires or deforestation. These species colonise the cleared areas, gradually creating a canopy. These pioneers provide shade, which allows shade-tolerant species to establish and diversify the ecosystem. The complex interactions between plants, animals, and microorganisms contribute to the restoration of biodiversity. Importantly, rainforest regeneration can be enhanced through conservation efforts, such as reforestation projects and sustainable land management practices. A study conducted on rainforests spanning three continents found that soil restoration takes approximately ten years to regenerate to its original status, whereas plant and animal biodiversity takes about sixty years, and overall biomass takes about one hundred and twenty years. This is mostly because of a process known as "secondary succession," in which adjacent patches of flora and woodland serve as a catalyst for future growth. Certain tropical forests can

reach about 75 percent of their previous growth status in just 20 years if left alone.

6. **As temperate rainforests are located in middle latitudes where much of the planet's population is, many forests were cut down to make room for big cities such as London, New York City, Paris and Tokyo.**
The demand for urban expansion has led to the clearing of vast forested regions to make way for metropolises like these major cities. The transformation of these once-thriving ecosystems into urban landscapes has not only resulted in the loss of valuable biodiversity but has also contributed to environmental issues such as habitat fragmentation. Achieving a balance between human growth and the preservation of these essential forested regions requires conservation efforts and sustainable urban planning, as temperate rainforests are valued for their ecological significance and the services they provide.

7. **Habitat fragmentation in forests refers to the process by which large, continuous forest areas are broken into smaller, isolated patches due to human activities such as urban development, agriculture, and road construction.**
This process destroys habitats and disrupts the natural connectivity of ecosystems, impacting both plants and animals. This separation disrupts ecosystems by reducing the size of available habitats and isolating wildlife populations, making it harder for animals to find food, mates, and safe migration routes.

Consequently, species that depend on moving long distances may face population decline and increased vulnerability to extinction. Fragmented habitats also reduce biodiversity, as smaller populations are more vulnerable to extinction from disease, natural disasters, and genetic inbreeding.

8. **While some forest fires are natural and play a role in ecosystem health, human-induced wildfires and uncontrolled fires can destroy vast forest areas, harm biodiversity, and release significant amounts of carbon into the atmosphere.**
Forest fires, also known as wildfires or bushfires, are uncontrolled and often intense blazes that spread rapidly through forested areas, grasslands, and other vegetation. These fires can be ignited by lightning, human activities, or a combination of both. While some wildfires are a natural part of certain ecosystems, the increasing frequency and intensity of human-induced fires pose significant threats. Strong winds, elevated temperatures, and dry conditions can accelerate the spread of these fires, leading to devastating social, economic, and environmental consequences.

9. **Slash-and-burn agriculture is an ancient farming method that involves clearing land by cutting and burning vegetation.**
Typically practised in tropical and subtropical regions, this technique involves farmers slashing and burning the vegetation to create nutrient-rich ash, which is used as a crop fertiliser. While slash-and-burn agriculture

can be sustainable when practised in small-scale and traditional settings, its widespread and uncontrolled application poses environmental challenges. The method can lead to deforestation, loss of biodiversity, and soil degradation as the protective forest cover is removed. Additionally, the carbon released during burning contributes to greenhouse gas emissions. Efforts to make this practice more sustainable involve implementing controlled burning, rotating cultivation areas, and incorporating agroforestry techniques to minimise its ecological impact.

10. **One of the most devastating wildfires in modern history occurred in Australia, starting in late 2019 and continuing into 2020. The scale of the disaster was staggering, with over 18 million acres of land burned, an estimated 3 billion animals affected, and tragic loss of human lives.** Dubbed the "Black Summer," the bushfires scorched vast expanses of land, predominantly in the southeastern parts of the country. These fires were fueled by prolonged drought, very high temperatures, and strong winds. They destroyed homes, displaced communities, and caused unprecedented ecological damage. For weeks, major cities, including Sydney and Melbourne, were blanketed in thick smoke. The severity of the Black Summer highlighted how dramatically climate change affects the frequency and intensity of wildfires. The devastating incident serves as a sobering reminder of the far-reaching consequences of extreme weather events and the importance of addressing environmental challenges on a global scale.

11. **Forest pathology is a specialised branch of plant pathology dedicated to studying diseases that affect trees and forest ecosystems.**
    Forest pathologists investigate various biotic and abiotic factors that can lead to tree diseases, including fungi, bacteria, viruses, insects, and environmental stresses like drought or waterlogging. These scientists' goals are to understand the patterns of forest health, identify pathogens responsible for diseases, and develop strategies for disease management and prevention. Forest pathology research contributes to sustainable forest management practices, helping reduce the impact of diseases on timber production and biodiversity.

12. **Illegal logging is a widespread and destructive practice involving the unauthorised harvesting, transportation, and trade of timber.**
    This illicit activity poses a severe threat to global forests, contributing to deforestation, loss of biodiversity, and environmental degradation. Often driven by the demand for valuable timber, illegal logging occurs in protected areas and indigenous territories, undermining conservation efforts and community livelihoods. The environmental consequences extend beyond the immediate deforestation, affecting ecosystems and water quality and contributing to climate change. Additionally, illegal logging often involves corruption and organised crime, making the social and economic challenges associated with this issue worse. Efforts to combat illegal logging include:
    - Strengthening law enforcement.

- Implementing sustainable forestry practices.
- Promoting responsible consumer choices to reduce the market for timber that is sourced illegally.

13. **By clearing forests, humans are often disrupting habitats and pushing wildlife closer to human settlements, which increases the likelihood of zoonotic diseases (those that can transfer from animals to humans), like Ebola and certain coronaviruses.**

When natural habitats shrink, animals like bats, rodents, or primates, which may carry pathogens, are more likely to interact with people and domestic animals, creating opportunities for disease transmission. For example, in the Amazon rainforest, deforestation has been linked to an increase in malaria cases. As trees are removed and land is cleared, stagnant pools of water often form in the exposed areas, creating ideal breeding grounds for mosquitoes, especially the *Anopheles* mosquitoes that transmit malaria. Studies have shown that these cleared areas are significantly more likely to harbour mosquito populations, increasing the risk of malaria transmission among local communities and settlers moving into newly deforested lands.

The spread of the Nipah virus in Southeast Asia provides another example of how deforestation can drive disease transmission. As forests in Malaysia and Bangladesh were cleared, fruit bats lost significant portions of their habitat and began foraging closer to farms and urban areas. In Malaysia, bats were found roosting in fruit trees near pig farms, leading to the transmission of Nipah virus to pigs and, subsequently, to farm workers who handled them. This cross-species transmission caused severe outbreaks, with high mortality rates in humans. The Nipah virus remains a public health threat, highlighting how deforestation-fueled habitat loss can create new pathways for diseases to jump from wildlife to humans.

14. **The Japanese practice of *shinrin-yoku*, or forest bathing, involves spending time immersed in a forest environment and has measurable health benefits. Some hospitals in Japan offer forest therapy as a treatment for patients dealing with stress-related illnesses.**

Unlike hiking, forest bathing is not about covering distance but rather about being fully present and mindful in nature, engaging all the senses to experience the sights, sounds, and smells of the forest. Studies show that *shinrin-yoku* has measurable health benefits: it reduces levels of cortisol (a stress hormone), lowers blood pressure, boosts our immune system, and improves mood. Forest air contains phytoncides, organic compounds released by trees, which have been shown to enhance immune function by increasing the activity of natural killer cells that help the body fight infections and even some types of cancer. Hospitals in Japan and South Korea have begun incorporating forest therapy into treatment for stress-related conditions, and public health officials worldwide are recognising its value as a natural form of preventive care. With urbanisation on the rise and more people facing high stress levels, *shinrin-yoku* offers a simple, accessible way to reconnect with nature and improve well-being.

Deserts

Types of deserts

1. **Deserts are defined by their low levels of precipitation, typically receiving less than 250 millimetres (10 inches) of rain annually.**
   A desert is a vast and typically barren expanse of land characterised by its extreme aridity, receiving minimal precipitation that is often insufficient to support most forms of life. They make up about a third of the Earth's surface, which includes the polar regions. Despite their seemingly inhospitable conditions, deserts are diverse and can be found across the globe, taking on various forms, such as subtropical deserts, cold deserts, coastal deserts, and rain shadow deserts. These regions exhibit unique ecosystems and are home to specialised plant and animal adaptations that enable survival in the challenging environment, including water-conserving features and temperature-regulating behaviours.

2. **The hottest deserts on Earth are called subtropical deserts, and they can be found in North and South America, Asia, Australia, and Africa.**
   These deserts are situated around 30 degrees latitude on both sides of the equator, where descending air

masses create high temperatures and minimal rainfall. As hot, moist air near the equator rises into the atmosphere, it cools and releases its moisture as heavy tropical rains. The resulting air mass shifts away from the equator as it becomes cooler and drier. As it approaches the tropics, the air descends and warms up again. The descending air restricts the formation of clouds, which causes the minimal rainfall in these deserts. The distinctive feature of subtropical deserts is their ability to adapt to harsh conditions, as they contain specialised flora and fauna equipped to endure the relentless sun and scarce water.

3. **The Sahara Desert, a type of subtropical desert and the largest hot desert in the world, is spread across North Africa, covering over 9 million square kilometres.**
The Sahara, known for its breathtaking sand dunes that can tower over 180 metres, experiences scorching daytime temperatures which often exceed 50 degrees Celsius (122 degrees Fahrenheit). Despite its seemingly inhospitable conditions, the Sahara is not a lifeless wasteland. The desert hosts a variety of resilient plant and animal species, including the iconic date palm, acacia trees, and the fennec fox. Additionally, ancient cultures such as the Berbers have thrived in the Sahara, leaving behind remarkable historical sites like the city of Timbuktu.

4. **Coastal deserts are unique ecosystems situated along the boundaries of continents, where arid conditions meet the sea.**

Coastal deserts are primarily located on the western edges of continental land masses in regions where cold currents approach the land or cold water upwellings rise from the ocean depths. The Namib Desert in Africa and the Atacama Desert in South America are two examples of these deserts. A unique feature of coastal deserts is the formation of heavy fog, formed by warm air drifting over the ocean's cool water and condensing to create a layer of fog. Coastal deserts often showcase remarkable biodiversity, with plant and animal species adapted to the specific challenges of these environments.

5. **The Atacama Desert, located in South America along the western coast of Chile, is renowned as one of the driest places on Earth. Some weather stations have never received a single drop of rain.**

The Atacama Desert stretches more than 1,000 kilometres (600 miles) along the Pacific Ocean, with the Pacific Ocean to the west and the Andes Mountains to the east. The desert's extreme aridity is due to its unique geographic and atmospheric conditions, including the cold Humboldt Current that flows just offshore and the rain shadow effect from the Andes. Rainfall is nearly non-existent in some parts of the Atacama, with some areas recording no measurable rainfall for decades. Despite its harsh conditions, the Atacama is not devoid of life; microorganisms, adapted plants, and animals like the vicuña and the Andean flamingo can be found here. The Atacama is also well-known for its abundant mineral resources, including the world's largest lithium reserves and extensive copper deposits.

6. **The Namib Desert, situated along the Atlantic coast of southern Africa, is recognised as one of the oldest and driest deserts in the world, with some areas experiencing less than 2 millimetres of rainfall annually.**
Spanning across Namibia, Angola, and South Africa, Namib's arid landscape is shaped by towering sand dunes that rise to staggering heights, including the famous Dune 45 and Big Daddy, which are 170 metres (557 feet) and 325 metres (1,066 feet) respectively. These dunes, formed by wind-blown sand over millions of years, create mesmerising patterns and provide a habitat for unique desert-adapted species such as the endemic Namib Desert beetle and the Welwitschia mirabilis, a long-lived plant species. Despite its harsh conditions, the Namib supports a surprising array of life, including desert-adapted wildlife like oryx, ostriches, and brown hyenas. Additionally, Namib's coastal fog, known as the "fog desert," sustains plant life in the form of lichens and provides a vital water source for desert-dwelling creatures.

7. **Cold deserts, also known as temperate or polar deserts, are unique ecosystems characterised by extreme cold temperatures and sparse vegetation.**
Cold deserts experience harsh winters with temperatures dropping below freezing, making it difficult for plants and animals to survive. One of the largest cold deserts is the Gobi Desert in Asia, which experiences frigid winters and scorching summers.

Another example is the Antarctic Desert, the largest desert on Earth, where temperatures can plummet to as low as -80 degrees Celsius (-112 degrees Fahrenheit). Despite the challenging conditions, cold deserts are home to resilient flora and fauna adapted to survive in extreme cold, including hardy shrubs, lichens, and animals like the Arctic fox.

8. **The Antarctic Desert, encompassing the continent of Antarctica, is the largest desert on Earth, characterised by its extreme cold, high winds, and low precipitation.**

   Despite being covered by vast ice sheets, Antarctica is categorised as a desert because it receives very little precipitation each year, which is primarily in the form of snow. The continent experiences freezing temperatures, with the lowest ever recorded being -89.2 degrees Celsius (-128.6 degrees Fahrenheit). Despite its harsh conditions, Antarctica is home to a variety of life forms, including penguins, seals, and various species of seabirds that thrive in the surrounding Southern Ocean. Scientific research conducted in Antarctica has provided valuable insights into climate change, glaciology, and astrobiology, making it a crucial area for global scientific exploration.

9. **Rain shadow deserts form on the leeward (sheltered) side of mountain ranges, created by the "rain shadow effect", where wet air masses release their moisture as they rise over the windward side of mountains, leaving dry air to descend and warm on the leeward side.**

This descending air warms adiabatically, reducing its relative humidity and inhibiting cloud formation and precipitation. As a result, rain shadow deserts like the Mojave Desert in the United States and the Patagonian Desert in South America experience minimal rainfall and arid conditions. These deserts are characterised by their unique ecosystems, adapted to survive with scarce water resources, and often feature distinctive landforms such as basins, valleys, and barren landscapes.

10. **Death Valley, located in Eastern California, USA, is the hottest and driest place in North America. It is part of the Mojave Desert and holds the record for the highest reliably reported air temperature on Earth, reaching 56.7 degrees Celsius (134 degrees Fahrenheit) in 1913.**

    This arid landscape features vast salt flats, towering sand dunes, and rugged mountains, including Telescope Peak, which rises to over 3,300 metres (11,000 feet) above sea level. Despite its harsh conditions, Death Valley supports a surprising diversity of life, including desert-adapted plants and animals such as the iconic Joshua tree and desert bighorn sheep. The valley's unique geology and extreme climate make it a popular destination for tourists, photographers, and adventurers seeking to explore one of the most inhospitable landscapes in the world.

11. **Mars is the only other planet in the Solar System besides Earth on which deserts have been identified.**

Deserts on Mars, known as Martian deserts, cover vast expanses of the planet's surface and are characterised by their arid, dusty landscapes. These deserts have rocky terrain, sand dunes, and expansive plains scattered with impact craters. Martian deserts experience extremely cold temperatures due to the planet's thin atmosphere, with average surface temperatures around -60 degrees Celsius (-76 degrees Fahrenheit). The largest desert on Mars is the vast expanse of sand known as the "grandest dune field," which stretches for over 3,000 kilometres (1,864 miles) near the planet's north pole. Despite the harsh conditions, these deserts are of significant interest to scientists studying Mars' geology, climate history, and potential habitability, as they provide valuable insights into the planet's past and present conditions.

Ecosystems

1. **Cacti's ability to retain water in their fleshy stems is a crucial trait that enables them to endure prolonged periods of drought.**
   Additionally, cacti have evolved specialised photosynthesis pathways, such as CAM (Crassulacean Acid Metabolism), which enable them to minimise water loss by opening their stomata at night and storing carbon dioxide for daytime photosynthesis. Their reduced leaf surface area and thick, waxy cuticles further help to prevent water loss through transpiration. Some cacti also have shallow, widespread root systems that efficiently absorb water from infrequent rainfall

events. These adaptations collectively allow cacti to thrive in some of the world's harshest desert environments, where water is scarce and temperatures are extreme.

2. **The Atacama Desert, one of the driest places on Earth, features lomas, areas in which fog condenses against mountain slopes near the sea and creates "fog oases" with an abundant biodiversity of plant and animal species.**
Plant coverage varies greatly, from coverage as high as 50% in the foggiest areas to a near-total absence of plant life above the fog line. These coastal desert oases rely on dense coastal fog, which provides essential moisture for plant life. Lomas ecosystems are typically located in valleys and depressions along the western edge of the Atacama Desert, where the fog condenses and provides a regular water source for specialised plant species. These fog-dependent plants, known as "fog vegetation," include species like lomas shrubs, cacti, and bromeliads. Lomas ecosystems support a variety of wildlife, including birds, insects, and small mammals, which rely on the vegetation for food and shelter. The unique adaptation of lomas plants to utilise fog moisture makes them essential components of the fragile ecosystems found in the Atacama Desert.

3. **Many desert animals, including rodents such as kangaroo rats, reptiles such as snakes, and insects such as beetles and scorpions, are primarily active during the cooler nighttime hours to avoid the day's extreme heat.**

They emerge at night to forage for food when temperatures are cooler. By avoiding the intense heat during the day, these animals can conserve energy and reduce water loss through evaporative cooling. Nocturnal behaviour also helps desert animals avoid predators that are more active during the day, as well as take advantage of lower temperatures for hunting or foraging. Additionally, some desert plants bloom at night, attracting nocturnal pollinators like bats and moths.

4. **Desert animals often have camouflage adaptations to blend in with their surroundings, helping them evade predators or ambush prey.**
Camouflage allows desert animals to blend seamlessly into their surroundings, protecting them from predators and aiding in hunting or ambush strategies. Examples of desert animals with exceptional camouflage include the sand-coloured fur of the desert fox, which helps it blend into the sandy desert terrain, and the colouration of the desert horned lizard, which matches the colour and texture of desert sands, allowing it to avoid detection by predators. Some desert insects, such as the desert grasshopper, have evolved body shapes and colours that mimic the texture and colouration of desert vegetation, providing them with camouflage against both predators and prey.

5. **Several desert animals, such as desert tortoises and burrowing owls, have adapted to dig burrows to escape the day's extreme heat and seek shelter from predators.**

Many desert species, including rodents like kangaroo rats, reptiles like desert tortoises, and insects like ants and termites, have evolved specialised behaviours and anatomical features for digging and living in burrows. Burrowing provides these animals with protection from the intense heat and predators, as well as insulation from temperature fluctuations. Additionally, burrows can serve as refuges during periods of extreme weather, such as sandstorms or flash floods. Some desert animals, like the fennec fox, also use burrows as dens for rearing their young. By utilising burrows, desert animals can create microhabitats with more stable environmental conditions, increasing their chances of survival.

6. **Desert animals like the camel have evolved physiological adaptations to tolerate high temperatures. Camels can conserve water and regulate body temperature efficiently, allowing them to survive in scorching desert environments.**

Heat tolerance is a crucial adaptation for desert animals to survive in the extreme temperatures of their arid habitats. Camels have evolved a range of physiological and behavioural adaptations to cope with high temperatures. Their unique metabolism allows them to go for long periods without drinking water, as they can rehydrate rapidly and store water in their bodies for extended periods. Additionally, camels have specialised nasal passages that allow them to reabsorb moisture from their exhaled breath, minimising water loss. Their ability to tolerate high temperatures makes camels well-suited for traversing deserts' hot, arid landscapes.

7. **Desert plants often have seeds with dormancy mechanisms that allow them to remain viable until conditions for germination are favourable, such as after a rare rainfall event.**
Seed dormancy is a critical adaptation many plant species employ in desert ecosystems to survive the unpredictable and harsh conditions of arid environments. These mechanisms can include physical barriers to germination, such as hard seed coats, or physiological dormancy, where seeds require specific environmental cues, such as moisture or temperature fluctuations, to break dormancy and initiate germination. By remaining dormant until conditions improve, desert plant seeds can avoid germinating during unfavourable periods, such as droughts or extreme temperatures, and increase their chances of successful establishment when moisture becomes available. This adaptation ensures the long-term survival and persistence of plant populations in desert habitats by enabling them to effectively time their germination to coincide with periods of optimal environmental conditions.

8. **Ephemeral plants in deserts have adapted to germinate and complete their life cycles rapidly in response to brief periods of rainfall. These plants capitalise on short-lived moisture to reproduce before conditions become too harsh again.**

These plants have evolved specialised adaptations to germinate quickly and grow, flower, and set seed within a short time frame following rainfall events in arid environments. Ephemeral plants often have small seeds that remain dormant in the soil until triggered by sufficient moisture, such as the rare desert rain. Once germinated, they grow rapidly, taking advantage of the temporary increase in soil moisture and nutrients. Examples of ephemeral plants in deserts include species like desert annuals, which produce seeds that lie dormant for years until favourable conditions arise, and opportunistic species like tumbleweeds, which disperse their seeds when triggered by environmental cues like wind or disturbance. The ability of ephemeral plants to respond rapidly to favourable conditions allows them to exploit short-lived opportunities for growth and reproduction in their harsh and unpredictable environment.

9. **Some desert animals, such as certain species of reptiles, snails, and frogs, undergo estivation, a state of dormancy similar to hibernation, during periods of extreme heat and drought. This adaptation helps them conserve energy and water.**
Similar to hibernation, estivation involves entering a state of dormancy to conserve energy and water during unfavourable environmental conditions. Desert animals that estivate typically retreat to underground burrows or seek out sheltered locations to escape the intense heat and dehydration of the desert environment. During estivation, these animals lower their metabolic rate and

become inactive, reducing their water loss and energy expenditure until conditions improve. Estivation allows desert animals to endure prolonged periods of harsh environmental conditions.

10. **Symbiotic relationships are common in desert ecosystems, where organisms rely on mutually beneficial interactions to enhance their chances of survival in harsh environments.**

    The yucca plant and yucca moth's symbiotic relationship is a well-known example of symbiosis in deserts. The female yucca moth lays her eggs inside the flowers of the yucca plant, pollinating the flowers in the process. The developing larvae feed on some of the yucca seeds but leave enough for the plant's reproduction. This mutualistic relationship benefits both partners: the yucca moth gains a suitable place to lay its eggs, while the yucca plant ensures its pollination and subsequent seed dispersal.

11. **Predators are essential to the functioning of desert ecosystems as they control prey populations and contribute to the overall balance of these hostile environments.**

    Examples of predators in deserts include carnivorous mammals like foxes, coyotes, and big cats such as mountain lions, as well as reptiles like snakes and birds of prey such as eagles and hawks. These predators have specialised hunting techniques suited to desert conditions, such as stealthy stalking, ambush tactics, and keen senses to locate prey in vast, open landscapes. Additionally, desert predators often have physiological adaptations to cope with extreme temperatures and conserve water.

12. **Some desert plants, like mesquite trees and creosote bushes, have deep root systems that enable them to tap into groundwater reserves, providing them with a consistent supply of water during dry conditions and enabling them to withstand extended droughts.**
Deep-rooted plants are crucial components of desert ecosystems, as they play a vital role in accessing groundwater and stabilising soil in arid environments. By reaching deep into the soil, these plants can access moisture that is unavailable to shallow-rooted species, allowing them to thrive in water-limited environments. Their enormous root systems also help to prevent desertification and soil erosion by anchoring the soil and lowering the risk of wind and water erosion. Mesquite trees, characterised by their compound leaves and long pods, and creosote bushes, known for their resinous leaves and yellow flowers, are both highly adaptable to various desert habitats, from sandy dunes to rocky slopes.

13. **The saguaro cactus has evolved physiological mechanisms to tolerate drought, such as temporarily storing water in its tissues during dry periods.**
Known for its towering stature and distinctive silhouette, the saguaro cactus is the largest cactus species in the United States, capable of reaching heights of up to 12 metres (40 feet) and living for over 150

years. Saguaros have a unique and slow growth rate, with young cacti taking up to 75 years to develop their first arm-like branches. These arms provide additional surface area for photosynthesis and reproductive structures. One of the most remarkable adaptations of this cactus is its ability to store a lot of water in its accordion-like pleats and spongy tissue, allowing it to survive prolonged periods of drought. The saguaro cactus is also essential for providing habitat for many desert animals, including birds, bats, and insects, which rely on the cactus for food, shelter, and nesting sites. The saguaro cactus is also culturally significant to indigenous peoples of the Southwest, who have utilised its fruits, seeds, and wood for various purposes for centuries.

14. **The Saharan silver ant is a species of ant that has adapted to thrive in the extreme heat of the Sahara Desert. Known for its silvery sheen and impressive heat tolerance, the Saharan silver ant is one of the few animal species known to withstand temperatures exceeding 50 degrees Celsius (122 degrees Fahrenheit).**
These ants are active foragers, scavenging for food across the hot desert sands, and their unique adaptation lies in their ability to regulate their body temperature during foraging trips. When venturing out into the scorching desert, Saharan silver ants navigate quickly, moving at speeds of up to one metre per second while simultaneously minimising their exposure to the intense heat. Their long, slender legs minimise contact with the hot sand, and their reflective silver hairs help dissipate

heat from their bodies. Additionally, the Saharan silver ant has a specialised glandular system that produces heat-shock proteins, which protect their cells from damage caused by extreme temperatures.

15. **Desert ecosystems are characterised by sparse vegetation, with plants spaced apart to minimise competition for limited water and nutrients. This adaptation allows desert plants to access resources more efficiently.**

    The limited availability of water and nutrients in deserts creates challenging conditions for plant growth, leading to low plant density and diversity across these landscapes. Sparse vegetation in deserts is typically composed of adapted plant species such as succulents, xerophytes, and drought-tolerant shrubs, which have evolved specialised adaptations to survive in arid conditions. These adaptations include reduced leaf surface area to minimise water loss through transpiration, deep root systems to access groundwater reserves, and the ability to store water in their tissues. The spacing of desert plants is also often optimised to minimise competition for limited resources such as water and sunlight. Despite the sparse vegetation cover, desert ecosystems are still home to a variety of plant species that play crucial roles in supporting desert wildlife.

16. **Herbivores in desert ecosystems, such as desert bighorn sheep and desert kangaroos, have specialised digestive systems that enable them**

to draw moisture and nutrients from tough, fibrous plant material like cacti.

They possess complex stomachs with multiple chambers, including a fermentation chamber where plant material is broken down by symbiotic bacteria, enabling them to derive maximum nutrition from their diet. Additionally, bighorn sheep are well adapted to conserve water, with highly concentrated urine and the ability to extract moisture from food sources, reducing their dependence on free-standing water sources. Similarly, desert kangaroos, like the red kangaroo found in Australian deserts, have adapted to survive in arid environments by primarily feeding on tough, drought-resistant grasses and shrubs. They can also conserve water through efficient kidney function and behavioural adaptations such as minimising activity during the hottest parts of the day to reduce water loss through sweating. Both bighorn sheep and desert kangaroos exhibit nomadic behaviour, moving in search of food and water as resources become scarce in their arid habitats.

17. **Found in desert regions of North Africa and the Middle East, the Deathstalker is one of the most venomous scorpions in the world. Despite its small size, its venom can be lethal to humans, though fatalities are rare.**

    Measuring around 8 centimetres in length, the Deathstalker is relatively small but possesses one of the most potent venoms among scorpions. The venom of the Deathstalker contains a potent cocktail of neurotoxins, which can cause severe pain, muscle spasms, paralysis, and even death in extreme cases, particularly for children or individuals with compromised immune systems. Despite its fearsome reputation, the Deathstalker is vital to its ecosystem, as it regulates insect populations. This scorpion is nocturnal, seeking shelter during the day and emerging at night to hunt for prey, primarily consisting of insects, spiders, and other small arthropods. Its venom is also being studied for potential medical applications, particularly in cancer research, where certain components have shown promise in targeting cancer cells.

18. **The desert tortoise is adapted to the deserts of North America and can survive for long periods without water by storing it in their bladders. They also dig burrows to escape extreme temperatures and predators.**
The Mojave and Sonoran Deserts in North America are home to the desert tortoise species. One of their most remarkable adaptations is their ability to keep water stored in their bladders, which allows them to survive for long periods without drinking. They also have thick, scaly skin and sturdy limbs that enable them to dig burrows for shelter from extreme temperatures and predators. Desert tortoises are herbivores, feeding primarily on grasses, herbs, and cacti, and they play a crucial role in desert ecosystems by dispersing seeds and creating burrows that provide habitat for other desert species. Unfortunately, desert tortoise populations have declined significantly due to habitat loss, fragmentation, and threats such as habitat

destruction, disease, and predation. Conservation measures are underway to preserve this iconic desert species and guarantee its survival for future generations.

19. **The thorny devil is a unique lizard species native to Australian deserts. It has a spiny body and a false head on the back of its neck, which it presents to predators when threatened. It also has specialised grooves on its skin that channel water from rain or dew to its mouth.**

The thorny devil has spiky, textured skin covered in conical scales that resemble thorns, providing effective camouflage against predators and blending in with its surroundings of sandy and rocky habitats. Despite its fearsome appearance, it is a relatively small lizard, growing up to about 20 centimetres in length. One of its most fascinating adaptations is its ability to collect water from the surface of its skin using a network of tiny grooves. During rainfall or when walking through dew-covered vegetation, the thorny devil channels water from its body to its mouth through capillary action, allowing it to stay hydrated in an environment where water sources are scarce. Additionally, this species feeds primarily on ants, using its specialised tongue to capture them.

20. **Found in North American deserts, kangaroo rats have specialised kidneys that enable them to efficiently obtain water from their food, allowing them to survive without drinking free water for extended periods.**

Despite their name, kangaroo rats are not closely related to kangaroos but are so named due to their large hind legs, which they use for jumping, similar to kangaroos. These rodents have adapted remarkably to their arid habitats, possessing several unique features that enable them to thrive in environments with little access to food and water. One of their most notable adaptations is their ability to obtain most of their water requirements from metabolic processes, allowing them to survive without drinking free water for extended periods. They achieve this by efficiently metabolising the water produced during the digestion of seeds and by concentrating their urine to conserve water. Kangaroo rats are primarily nocturnal and have excellent hearing and agility, enabling them to evade predators such as owls, snakes, and foxes. They mainly feed on seeds, which they gather and store in cheek pouches to transport to their burrows for consumption. Kangaroo rats play a crucial role in desert ecosystems as seed dispersers, helping to maintain plant diversity and ecosystem stability.

21. **Desert winds, often accelerated by temperature differentials and topographic features, can reach high speeds due to the absence of vegetation and obstacles, leading to the formation of powerful windstorms.**
In the Sahara, strong windstorms, known as "Harmattan" in West Africa and "Sirocco" or "Khamsin" in North Africa, can generate gusts exceeding 100 kilometres per hour (62 miles per hour). These intense windstorms lift vast amounts of sand and dust into the

atmosphere, creating massive dust storms that can span thousands of kilometres. The Sahara's powerful dust storms are capable of lifting fine particles high into the atmosphere, where they can be carried across continents and even oceans. These airborne dust particles can travel thousands of kilometres, impacting air quality, influencing weather patterns, and contributing to the transport of nutrients and pollutants across vast distances.

22. **Welwitschia is a remarkable plant native to the harsh deserts of Namibia and Angola. It is often called a "living fossil" because its primitive features, such as its unique growth pattern and reproductive system, have remained largely unchanged for millions of years.**

The term "living fossil" is used to describe organisms that have survived with little evolutionary change while many related species have gone extinct. This plant's lineage dates back over 100 million years, and its unique characteristics have persisted since that time, offering scientists a direct window into the prehistoric past. Its ancient, unaltered form provides insights into early plant evolution and how some species can remain stable over vast geological timescales, adapting to extreme conditions without significant changes to their structure or biology. The plant consists of just two leaves that grow continuously throughout its lifespan, which can exceed 1,000 years. Tattered and weathered by time, these leaves spread out from a stout, woody base, creating an unusual appearance that defies typical plant morphology. Welwitschia's survival in such an arid environment is a testament to its extraordinary adaptations, including deep roots and the ability to absorb moisture from fog, making it one of the planet's most resilient species.

Human survival in deserts

1. **The ancient city of Petra in Jordan, situated in the Arabian Desert, showcases the architectural achievements of desert civilisations, including rock-cut structures and water conservation systems.**
   Founded as the capital of the Nabataean Kingdom around the 4th century BCE, Petra flourished as a major trade hub strategically positioned along the caravan routes that linked the Arabian Peninsula, Egypt, and the Mediterranean region. Its unique structures, including the iconic Treasury (Al-Khazneh), the Monastery (Ad Deir), and the Roman Theater, are carved directly into the rose-coloured sandstone cliffs, showcasing the advanced engineering and artistic skills of the Nabataeans. Petra's prosperity declined after it was annexed by the Roman Empire in 106 CE. It later fell into obscurity and was abandoned by the 7th century CE. Rediscovered by the Western world in the early 19th century, Petra has since become a UNESCO World Heritage Site and a symbol of Jordanian heritage, attracting visitors from around the globe to marvel at its ancient marvels and rich cultural heritage.

2. **Desertification, the process of land turning into desert, is a growing concern, often linked to human activities such as deforestation and improper agricultural practices.**

   Desertification is a complex process of land degradation that occurs in dryland ecosystems due to various factors, including climate change, unsustainable land management practices, overgrazing, deforestation, and inappropriate agricultural activities. It involves the gradual transformation of fertile, productive land into arid, barren desert landscapes, often resulting in the loss of biodiversity, soil fertility, and water resources.

3. **Despite the arid conditions, humans have developed techniques for agriculture in desert regions.**

   Agriculture in deserts faces difficulties due to the scarcity of water and extreme environmental conditions, yet humans have developed creative techniques to cultivate crops and sustain livelihoods. One notable technique is drip irrigation, which minimises evaporation and maximises water usage by delivering water straight to the roots of plants. Additionally, desert agriculture often relies on the cultivation of drought-resistant crops such as dates, figs, olives, and certain varieties of grains like millet and sorghum, which can thrive in dry conditions. Traditional practices such as terrace farming and contour ploughing help to conserve soil moisture and prevent erosion. Modern technological advancements, including greenhouses with controlled environments and hydroponic systems, further enhance agricultural productivity.

4. **Humans have established cities and infrastructure in desert regions, such as Las Vegas in the Mojave Desert or Dubai in the Arabian Desert. These cities often rely on advanced engineering and technology to handle problems with extreme temperatures and water scarcity.**

Deserts provide unique potential and challenges for urbanisation and infrastructure development due to variables such as population growth, economic development, and geographic location. Advanced engineering and technology are crucial in overcoming desert-specific challenges, including water scarcity and extreme temperatures. Smart irrigation techniques, wastewater recycling systems, and desalination plants are some of the innovative technologies used to meet urban populations' water needs and support infrastructure development. Moreover, desert cities often incorporate sustainable design principles, such as using reflective materials and green building techniques to mitigate heat island effects and reduce energy consumption. The strategic location of desert cities along trade routes and near natural resources has also fueled economic growth, leading to the construction of transportation networks, including highways, railways, and airports, to facilitate trade and connectivity. However, rapid urbanisation in deserts can also exacerbate environmental degradation, with issues such as habitat destruction, air and water pollution, and increased demand for resources posing significant challenges to sustainability.

5. **Deserts are sometimes used by military forces for training exercises and weapons testing due to their vast, uninhabited expanses and similarity to certain conflict zones. For example, the United States military conducts training exercises in the Mojave Desert at places like the National Training Center at Fort Irwin.**

   Deserts have long served as valuable environments for military training and testing, providing realistic conditions for preparing troops and testing equipment. Countries such as the United States, Israel, and Saudi Arabia utilise desert regions like the Mojave Desert in California, the Negev Desert in Israel, and the Rub' al Khali in Saudi Arabia for military exercises. These exercises often simulate combat scenarios in harsh desert conditions, including extreme temperatures, limited water sources, and rugged terrain, preparing soldiers for deployment to arid regions around the world. Desert environments also offer ideal settings for testing a variety of military equipment under extreme operational scenarios and weather conditions, such as vehicles, aircraft, and weapons systems. Additionally, desert training facilities provide opportunities for international military cooperation and joint exercises, fostering interoperability and collaboration among allied forces. While military training and testing in deserts are essential, environmental conservation efforts are increasingly emphasised to minimise the impact on fragile desert ecosystems and preserve their natural resources for future generations.

6. **Deserts are increasingly being used for renewable energy development, particularly for solar and wind power generation, due to their abundant sunlight and consistent wind patterns. The abundance of sunlight and available land make deserts ideal locations for large-scale solar and wind farms.**

   Solar energy projects, such as concentrated solar power (CSP) and photovoltaic (PV) solar farms, thrive in desert regions where clear skies and high solar irradiance levels maximise energy production. Countries like the United States, China, Morocco, and the United Arab Emirates have invested in large-scale solar power plants in deserts, such as the Ivanpah Solar Power Facility in the Mojave Desert and the Noor Complex Solar Power Plant in Morocco's Sahara Desert. These projects significantly reduce greenhouse gas emissions and increase renewable energy production. Similarly, wind farms harnessing desert winds have been established in regions like the Gobi Desert in China and the Mojave Desert in the United States, providing clean energy to local communities and supporting international efforts to make the shift to a sustainable energy future.

7. **Desert greening refers to the process of reclaiming and restoring arid and semi-arid desert lands through various ecological and agricultural techniques aimed at increasing vegetation cover and restoring ecosystem health.**

This approach involves implementing a combination of sustainable land management practices, such as reforestation, afforestation, agroforestry, and soil conservation measures, to improve soil fertility, enhance water retention, and promote biodiversity in desert environments. Techniques like micro catchment water harvesting, where small-scale structures capture and store rainfall runoff, and the use of drought-tolerant plant species help mitigate water scarcity and create microclimates, supporting plant growth.

8. **The Great Green Wall is an ambitious transcontinental initiative that aims to combat desertification and climate change in Africa by planting a wall of trees across the Sahel region.**
Spanning over 8,000 kilometres from Senegal in the west to Djibouti in the east, the Great Green Wall involves planting a belt of trees and vegetation to restore degraded lands, promote biodiversity, and improve the livelihoods of millions of people living in the Sahel. The initiative was started by the African Union in 2007 and aims to combat desertification, land degradation, and drought by restoring degraded lands and providing sustainable land management practices. By restoring ecosystems and enhancing natural resources, the Great Green Wall seeks to mitigate the impacts of climate change, increase food security, and create jobs in rural communities. While progress has been made in planting trees and implementing green infrastructure, challenges such as funding constraints, political instability, and coordination among multiple stakeholders still need to be addressed.

9. **Mirages are optical illusions that occur in desert environments due to the refraction of light as it passes through layers of air with different temperatures and densities near the surface. In deserts, where the air near the ground is often significantly hotter than the air above, mirages are a common phenomenon, particularly during the hottest parts of the day.**

   These illusions typically manifest as shimmering patches of light or distorted images that appear to be water, often on the horizon. The most common type of mirage in deserts is the "inferior mirage," where the hot air near the surface causes light to bend upward, creating the illusion of water or a reflective surface below. Mirages can be misleading for travellers, as they can give the impression of real bodies of water that are not actually present, leading to disorientation and dehydration if not recognised. However, mirages also have practical applications, such as aiding in navigation by indicating the presence of temperature gradients and atmospheric conditions.

10. **Sandstorms can bury everything in their path— including rocks, fields, and even towns. An old legend states that in 530 BCE, the Persian Emperor Cambyses II led an army of 50,000 troops to the Siwa Oasis in western Egypt. However, the army was completely engulfed by a massive sandstorm when they were about halfway there. Even now, archaeologists have**

spent years unsuccessfully searching for the "Lost Army of Cambyses" in the Sahara.

The exact fate of the army remains unknown, with theories ranging from succumbing to natural elements such as sandstorms and dehydration to being attacked and overwhelmed by local tribes. Despite numerous expeditions and searches over the centuries, the remains of the lost army have never been definitively located, adding to the intrigue and speculation surrounding the mystery.

11. **Despite the harsh conditions, marathons are held in deserts worldwide. The most famous is the Marathon des Sables in the Sahara Desert, covering over 150 miles in six stages.**

    The Marathon des Sables, often hailed as the "toughest footrace on Earth," is an intense multi-stage ultramarathon held annually in the searing heat of the Moroccan Sahara Desert. Founded in 1986 by Frenchman Patrick Bauer, who had previously completed a solo trek across the Sahara, the race was initially conceived as a way to share the profound experience of crossing the desert on foot. Today, it has grown into an iconic event, attracting over 1,000 participants from around the world each year. Spanning approximately 241 kilometres (150 miles) over six days, participants must carry their own supplies, battling through scorching temperatures that can exceed 50°C (122°F), relentless sand dunes, and rugged terrain. The Marathon des Sables is a true test of human limits, attracting adventurers from around the world.

Weather

Elements of weather

1. **Weather occurs throughout the Earth's atmosphere, which is the layer of gases that surrounds the planet.**
Weather encompasses a range of phenomena, including temperature, humidity, precipitation, wind, and atmospheric pressure. Weather occurs in the troposphere, the lowest layer of the atmosphere, which extends from the Earth's surface to an average height of between 8 to 15 kilometres (5 to 9 miles). Within the troposphere, the interaction of solar radiation with the Earth's surface drives atmospheric processes such as convection, evaporation, and condensation, leading to the formation of clouds, storms, and other weather events. Atmospheric circulation patterns, influenced by factors such as the uneven heating of Earth's surface, the rotation of the planet, and the distribution of land and water, further shape weather patterns on a regional and global scale.

2. **The energy that powers atmospheric processes is primarily derived from the Sun, which also determines Earth's weather patterns.**

The Sun is crucial in shaping Earth's weather patterns through various mechanisms. One significant impact is on air pressure. Solar radiation heats the Earth's surface unevenly, causing air masses to warm and rise in some areas while cooler air rushes in to fill the void, creating areas of high and low pressure. These pressure differences drive atmospheric circulation, influencing wind patterns and the movement of weather systems. Additionally, the Sun's energy directly affects temperature by warming the Earth's surface, leading to the formation of temperature gradients that drive atmospheric circulation. Moreover, solar energy powers the water cycle by evaporating water from oceans, lakes, and land surfaces, leading to differences in moisture levels across regions. These variations in temperature and moisture create the conditions necessary for the formation of clouds, precipitation, and other weather phenomena, ultimately shaping Earth's diverse climate systems.

3. **Weather on other planets in the solar system varies greatly due to differences in atmospheric composition, surface features, and distance from the Sun.**
For instance, Venus's thick atmosphere, primarily made of carbon dioxide, causes the planet to suffer severe temperatures of up to 470 degrees Celsius (870 degrees Fahrenheit). On Mars, dust storms can cover the entire planet, with temperatures ranging from -125 degrees Celsius (-195 degrees Fahrenheit) at the poles to 20 degrees Celsius (70 degrees Fahrenheit) near the equator. Saturn's weather events include massive

cyclones and lightning storms, while Uranus and Neptune have extremely icy conditions with strong winds and occasional storms.

4. **The composition of a planet's atmosphere plays a significant role in determining weather patterns. For instance, Earth's atmosphere primarily consists of nitrogen and oxygen, while other planets like Venus have atmospheres predominantly made up of carbon dioxide, which results in vastly different weather patterns.**

   Earth's atmosphere, predominantly composed of nitrogen and oxygen, supports a climate with moderate temperatures and varied weather conditions that are favourable to life. In contrast, planets like Venus, with atmospheres dominated by carbon dioxide and thick layers of sulfuric acid clouds, experience extreme heat due to the greenhouse effect. This results in surface temperatures hot enough to melt lead, forming a never-ending cycle of thick clouds and intense atmospheric pressure. Mars's thin atmosphere, primarily consisting of carbon dioxide, exhibits a starkly different climate characterised by cold temperatures and occasional dust storms. Atmospheric composition dictates a planet's ability to retain heat, regulate temperature gradients, and interact with solar radiation.

5. **Weather is not limited to planetary bodies; weather on the Sun, often referred to as solar weather, is dynamic and driven by complex**

processes occurring within its atmosphere, known as the solar corona.

The Sun experiences solar flares, coronal mass ejections (CMEs), and solar wind. Solar flares are intense bursts of radiation, often accompanied by the release of energetic particles, caused by the sudden release of magnetic energy stored in the Sun's atmosphere. Coronal mass ejections are massive plasma and magnetic field eruptions from the Sun's corona into space. They can influence space weather and trigger geomagnetic storms when they interact with Earth's magnetosphere. The entire solar system is impacted by solar wind, a continuous flow of charged particles that originates from the Sun and interacts with planetary magnetospheres, causing auroras and other atmospheric phenomena.

6. **Wind is the movement of air from areas of high pressure to areas of low pressure.**

It is caused as a result of the Sun unevenly heating the Earth's surface and is influenced by the planet's rotation, surface friction and geographic features like mountains and coastlines. Winds vary widely in speed and direction, from gentle breezes to powerful storms and cyclones. The larger the difference in pressure, the stronger the wind gets as it rapidly moves from high-pressure zones to low-pressure ones. They play a crucial role in shaping climate, distributing heat and moisture around the globe, dispersing pollutants, and facilitating the transport of seeds and pollen.

7. **The physical features of the Earth's surface play a crucial role in shaping weather patterns. Mountains, valleys, bodies of water, and other terrain features influence local climate by affecting temperature, precipitation, and wind patterns.**
Mountains act as barriers to air movement, forcing moist air to rise and cool, forming clouds and precipitation on the windward side. This process, known as orographic lifting, creates distinct climate zones, with wetter conditions on the windward side and drier conditions, called rain shadows, on the leeward side. Valleys can funnel winds, intensifying their speed and leading to localised wind patterns. Bodies of water, such as oceans and lakes, moderate temperatures by absorbing and releasing heat more slowly than land, leading to milder climates in coastal regions. Coastal areas also experience sea breezes, where cooler air from the water moves inland during the day, while land breezes occur at night as cooler air from land moves towards the warmer water.

Weather forecasting

1. **Weather forecasting has ancient roots, with early civilisations relying on observations of the sky, wind patterns, and animal behaviour to predict weather changes.**
One of the earliest documented methods dates back to around 650 BCE in Babylonia, where Babylonian astronomers tracked the movements of celestial bodies

to anticipate seasonal changes, while the Greeks observed the behaviour of animals and the appearance of clouds to forecast upcoming weather events. While these early methods lacked scientific rigour, they laid the foundation for understanding weather patterns and phenomena, paving the way for the development of more sophisticated forecasting techniques in later centuries.

2. **Modern weather forecasting began in 1935 with the invention of the electric telegraph.**
The electric telegraph enabled the transmission of messages over long distances at unprecedented speeds. Meteorologists quickly recognised the potential of the telegraph to improve weather forecasting by facilitating the rapid exchange of weather data between distant locations. Beginning in the mid-19th century, weather observation networks equipped with telegraph connections were established across Europe and North America. These networks allowed meteorologists to share real-time observations of temperature, pressure, wind direction, and other key weather parameters, enabling more comprehensive and accurate weather analysis. The telegraph also made it easier for the public to receive weather forecasts, allowing for the timely warning of approaching storms and other catastrophic weather events.

3. **The barometer has been used since the 19th century, allowing scientists to predict the weather based on measurements of pressure.**

The weight of the air molecules in the Earth's atmosphere exerts a force known as atmospheric pressure, which has a direct impact on weather patterns. A barometer is a glass tube filled with mercury or a similar liquid inverted in a dish of the same liquid. As atmospheric pressure changes, the level of the liquid in the tube rises or falls. Clear sky and fair weather are linked to high-pressure systems due to denser air descending and preventing clouds from forming. Conversely, low-pressure systems often bring cloudy or stormy weather, as rising warm air creates conditions favourable to precipitation. By monitoring changes in atmospheric pressure using a barometer, meteorologists can anticipate shifts in weather patterns and issue forecasts accordingly.

4. **The first-ever public weather forecast was published in The Times newspaper on 1st August 1861.**
In 1859, a violent storm off the coast of Anglesey, Wales, resulted in a shipwreck and the loss of 450 lives. This event inspired a British meteorologist, Robert FitzRoy, to begin developing charts that would allow weather forecasting. FitzRoy's forecast was based on observations collected by telegraph from a network of weather stations and ships equipped with barometers and thermometers. Although rudimentary by modern standards, FitzRoy's forecasts marked a significant step forward in the systematic prediction of weather patterns. His weather forecasting was initially for storm warnings for ships, which began in February 1861 with

the use of telegraph communication but was soon followed by daily publications in The Times.

5. **Numerical Weather Prediction (NWP) involves using mathematical models of the atmosphere to simulate and predict future weather conditions. These models consider various factors, including temperature, humidity, wind speed, and atmospheric pressure.**
The roots of NWP can be traced back to the work of Lewis Fry Richardson in the 1920s, who envisioned using numerical methods to solve the equations governing atmospheric motion. However, it wasn't until the development of electronic computers in the 1950s that Richardson's vision became feasible. In 1950, the Electronic Computer Project produced the first operational numerical weather forecast at the Institute for Advanced Study in Princeton, New Jersey. With the use of mathematical models of the atmosphere and oceans, NWP models can now simulate complex atmospheric processes in unprecedented detail and predict weather based on current meteorological conditions.

6. **Weather satellites, radiosondes, and weather radars are essential tools in weather forecasting, providing valuable data for monitoring and predicting atmospheric conditions.**
Satellites orbiting Earth capture imagery and data on cloud cover, temperature, humidity, and other atmospheric variables, providing a comprehensive view

of weather systems on a global scale. This information is crucial for tracking the development and movement of weather systems, including hurricanes and storms. Although satellite data has less accuracy and resolution, it has the benefit of global coverage. Radiosondes are weather instruments carried through the air by balloons, measuring temperature, humidity, pressure, and wind speed as they ascend through the atmosphere. Weather radars emit pulses of electromagnetic radiation to detect precipitation and measure its intensity and movement. This information is vital for monitoring severe weather events such as thunderstorms, tornadoes, and heavy rainfall, enabling timely warnings and advisories to be issued to the public. By integrating data from satellites, radiosondes, and weather radars, meteorologists can initialise and verify numerical weather prediction models, improving forecast accuracy.

7. **Weather drones, also known as atmospheric drones or unmanned aerial vehicles (UAVs) equipped with meteorological sensors, are emerging as valuable tools for weather monitoring and forecasting.**
Using satellites, radiosondes, and radars to gather data for weather forecasting leaves a gap in data collection in the lower atmosphere. To reduce this gap, weather drones were first explored for gathering data at those heights in the late 1990s. These drones can collect real-time data on atmospheric conditions such as temperature, pressure, humidity, wind speed, and air quality at various altitudes. By flying into regions

inaccessible to traditional weather observation methods, such as hurricanes, wildfires, or remote areas, weather drones can provide valuable insights into atmospheric processes and improve the accuracy of weather forecasts. Additionally, weather drones offer the flexibility to deploy rapidly in response to evolving weather events, allowing for targeted data collection and improving situational awareness for meteorologists and emergency responders.

8. **The instability of the atmosphere causes forecasting inaccuracy. Small changes to the starting conditions can cause significant variations in the forecasted outcomes.**

   Short-term forecasts, covering periods of up to three days, tend to be highly accurate, often predicting temperature, precipitation, and wind conditions with a high degree of precision. Medium-range forecasts, spanning up to a week, have also seen notable improvements in accuracy, particularly for large-scale weather patterns such as fronts and storm systems. However, forecasts beyond a week, known as long-range forecasts, are inherently more uncertain due to the chaotic nature of the atmosphere and the limitations of current modelling capabilities. Since the 1990s, ensemble forecasts have been used to help increase the accuracy and reliability of weather forecasts and quantify the large amount of uncertainty remaining in numerical predictions. This allows us to obtain useful results farther into the future than would otherwise be possible.

9. **Meteorologists often use ensemble forecasting to account for uncertainties in weather prediction. This helps determine forecast confidence by generating a range of probable outcomes by executing multiple simulations with slightly varying initial conditions.**

   Instead of relying on a single deterministic forecast, ensemble forecasting generates multiple forecasts, or ensemble members, by introducing slight variations in initial conditions or model parameters. By running multiple simulations with slight differences in starting conditions, ensemble forecasting produces a range of possible outcomes, known as the ensemble spread. Meteorologists analyse the spread and consistency among ensemble members to assess forecast uncertainty and confidence. Consistent trends among ensemble members provide higher confidence in the forecast, while divergent outcomes indicate greater uncertainty. Ensemble forecasting enables probabilistic forecasting, where the likelihood of different weather scenarios occurring is quantified.

10. **Accurate weather forecasting is crucial for public safety and disaster management, providing early warnings for severe weather events such as storms, heatwaves, and wildfires, and has saved many lives.**

    Weather forecasting is critical in public safety, providing advance warnings and crucial information about potentially hazardous weather conditions. Timely and accurate forecasts enable authorities to issue warnings and advisories for events such as hurricanes, tornadoes, severe thunderstorms, floods, and heatwaves, allowing individuals and communities to take action to protect life and property. For example, advance notice of an approaching hurricane can prompt evacuations and mobilise emergency response efforts, reducing the risk of casualties and facilitating efficient allocation of resources. Similarly, forecasts of extreme heat or cold can help vulnerable populations prepare by seeking shelter or adjusting their daily routines.

11. **Weather forecasting is essential in the agriculture and transportation sectors. It provides valuable information to support decision-making and mitigate risks.**

    In agriculture, weather forecasts help farmers plan planting and harvesting schedules, optimise irrigation and fertilisation practices, and manage pest and disease outbreaks. By anticipating weather patterns such as rainfall, temperature fluctuations, and frost events, farmers can make informed decisions to maximise crop yields and minimise losses. Additionally, forecasts of severe weather events such as droughts, floods, and hailstorms enable farmers to implement mitigation strategies, such as crop insurance or diversification of crops.

    In the transportation sector, weather forecasts are essential for planning and ensuring the safety and efficiency of travel by road, air, and sea. Transportation authorities use weather forecasts to monitor road conditions, deploy snow removal equipment during winter storms, and issue advisories to drivers, reducing the risk of accidents and improving traffic flow. Similarly, airlines use weather information to adjust flight routes, anticipate turbulence, and prepare for adverse weather conditions such as thunderstorms or fog, enhancing passenger safety and minimising delays.

The horrific 1977 Tenerife Airport disaster, where two planes collided on the runway as one began takeoff while another was still on the runway, killing 583 people, was partly due to the dense fog that had settled in. Visibility deteriorated to the point where the control tower and parts of the runway were obscured from the view of both aircraft. While it was not caused by weather alone, dense fog contributed to low visibility, complicating ground control efforts. Advanced forecasting and ground-level visibility monitoring are some of the many improvements that have been made since this disaster, preventing similar tragedies in the future.

12. **Climate change presents both opportunities and challenges for weather forecasting. Rising global temperatures can increase the frequency and intensity of weather events, requiring adjustments to forecasting techniques and infrastructure.**

Climate change can significantly impact weather forecasting by altering the frequency, intensity, and distribution of weather events. As the climate warms, changes in atmospheric circulation patterns, ocean currents, and temperature gradients can lead to shifts in weather patterns and increased variability. For instance, rising global temperatures may make extreme weather events like heat waves, torrential rainfall, and storms more likely. Additionally, variations in the frequency and strength of phenomena such as El Niño and La Niña can influence weather patterns on regional and global scales. These changes pose challenges for weather forecasting, as historical data may no longer accurately represent future conditions, and existing forecasting models may need to be adapted to account for shifting climate dynamics. Furthermore, climate change can also affect the predictability of weather events, with some studies suggesting increased uncertainty in future weather forecasts due to changing atmospheric conditions.

13. **Machine learning is revolutionising the field of weather forecasting. By analysing vast amounts of data, including historical weather patterns, satellite imagery, and real-time observations, sophisticated algorithms are learning to recognise complex weather patterns, meaning more accurate predictions can be made.**

    By using different types of machine learning, like recognising patterns or predicting numbers, these models can forecast things like temperature, rain, wind speed, and even storms. They can do this with increasing accuracy and can spot complex weather trends that humans might miss. This advancement is crucial for multiple sectors, from agriculture to transportation, as it enables better planning and preparedness for diverse weather conditions. Challenges such as poor data quality due to noisy or incomplete data or data containing errors persist. However, the potential for even more precise and timely forecasts continues to drive research and development in this field.

Extreme weather events and climate change

1. **In 1684, the Thames River in England froze solid for two months due to extreme weather.**
   The freezing of the River Thames in 1684 is a notable event in the history of London and meteorology. During the severe winter of 1683-1684, temperatures plummeted across Europe, leading to the freezing of water bodies, including the River Thames in England. The Thames froze considerably, with ice reportedly reaching a thickness of over a foot in some areas. This phenomenon allowed for various activities on the frozen river, such as ice skating, sledding, and even the construction of temporary markets and fairs. The freezing of the Thames was not an isolated event but rather part of a broader period of unusually cold winters known as the Little Ice Age, which lasted from the 14th to the 19th century. The freezing of the Thames served as a memorable historical event, documented in paintings and literature, and also provided valuable insight into past climatic conditions and the variability of weather patterns over time.

2. **Weather on other planets in the solar system has helped us understand how weather works. The Great Red Spot on Jupiter is one of the most iconic features in the solar system, a colossal storm that has been raging for centuries.**

   This massive anticyclonic storm is larger than Earth itself, with dimensions spanning approximately 15,000 miles (24,000 kilometres) in diameter. First observed in the 17th century, it is characterised by its distinctive deep red hue, believed to be caused by complex organic molecules interacting with solar radiation in Jupiter's atmosphere. Winds within the Great Red Spot can reach speeds of up to 400 miles per hour (640 kilometres per hour), creating intense turbulence and a swirling vortex of gas. Despite its long duration, the storm's precise mechanisms and the reasons for its reddish colouration remain subjects of scientific inquiry.

3. **In Kerala, India, in 2001, there was a blood-red rain event where heavy downpours of red-coloured rain occurred.**

   Beginning in July 2001, Kerala experienced episodes of coloured rainfall, ranging from yellow to red. Analysis of the coloured rainwater revealed the presence of microscopic red particles, leading to speculation about extraterrestrial origins, such as comet dust or microbial life from space. However, subsequent investigations suggested that the red particles were likely airborne spores of a terrestrial algae species called Trentepohlia annulata, which grows in moist conditions. These spores were lifted into the atmosphere by strong winds and carried over long distances before precipitating with rainfall. The red rainfall event in Kerala provided valuable insights into atmospheric transport mechanisms.

4. **One of the most infamous tornadoes in recorded history is the Tri-State Tornado of 18th March 1925 in the United States. It has the longest continuous path length of any tornado, stretching approximately 219 miles (352 kilometres), with estimated wind speeds exceeding 300 miles per hour (480 kilometres per hour).**

Tornadoes are highly destructive and rapidly rotating columns of air that originate from thunderstorms high in the sky and extend all the way to the Earth. These violent phenomena typically form when warm, moist air clashes with cold, dry air, creating instability in the atmosphere. The United States experiences the highest number of tornadoes annually, with an average of over 1,000 tornadoes recorded each year. Tornadoes vary in size, shape, and intensity, with wind speeds ranging from relatively mild to over 300 miles per hour, making them one of Earth's most powerful natural forces. Tornadoes can cause widespread devastation, demolishing buildings, uprooting trees, and tossing vehicles with ease. They are often accompanied by severe weather conditions such as hail, lightning, and heavy rain. Despite advances in meteorological technology and forecasting, tornadoes remain unpredictable, striking suddenly and leaving little time for preparation or evacuation.

5. **A region in Central America known as "Tornado Alley" is well-known for having a high frequency of tornadoes.**
Stretching primarily across parts of Texas, Oklahoma, Kansas, and Nebraska, this area experiences a convergence of climatic conditions that make it particularly susceptible to tornado formation. When cooler, drier air from the Rocky Mountains comes into contact with warm, humid air from the Gulf of Mexico, a favourable environment is created for the formation of intense thunderstorms and tornadoes. The unique geography of Tornado Alley, with its flat terrain and absence of geographical barriers such as mountains, allows these storms to travel long distances, often causing widespread damage along their paths. Tornadoes in Tornado Alley can vary in size and intensity, from relatively weak to catastrophic tornadoes with wind speeds exceeding 200 miles per hour. As a result of the frequency and intensity of tornado activity, residents of Tornado Alley have developed sophisticated warning systems and safety protocols to lower the risks associated with these powerful storms.

6. **Hailstones can vary widely in size, from tiny pellets to large chunks several inches in diameter. The largest hailstone ever recorded fell during a severe thunderstorm in Vivian, South Dakota, United States, on 23rd July 2010, with a diameter of 20.3 centimetres (8 inches).**
Hailstorms typically occur within thunderstorms, particularly those with strong updrafts that have the ability to lift raindrops into extremely cold areas of the

atmosphere, where they freeze and build up additional layers of ice. Hailstones can vary widely in size, from small pellets to large chunks several inches in diameter, depending on factors such as the strength of the updrafts and the duration of the storm. In areas where hailstorms occur frequently, crops, cars, buildings, and other property are particularly vulnerable to severe damage. In addition to property damage, hailstorms can threaten human safety, especially if individuals are caught outdoors without adequate shelter.

7. **Urban heat islands are areas where urban or metropolitan areas are significantly warmer than their rural surroundings due to human activities. During heat waves, these urban heat islands can exacerbate temperatures, leading to increased heat-related illnesses and energy demands.**

Urban heat islands are areas within cities or metropolitan areas that, as a result of heat-absorbing and heat-retaining human activity like roads, buildings, and vehicles, are significantly warmer than the surrounding rural areas. Urban areas typically have surfaces such as concrete and asphalt that absorb and retain heat, along with a scarcity of vegetation to provide cooling through shade and evapotranspiration. As a result, urban heat islands can experience temperature differences of several degrees Celsius compared to nearby rural areas, particularly during hot weather conditions. The elevated temperatures in urban heat islands can exacerbate heat-related illnesses such as heat exhaustion and heatstroke, increase energy

consumption for cooling, and worsen air quality by enhancing the formation of pollutants. Efforts to mitigate urban heat islands include:
- Increasing green spaces.
- Implementing cool roofing and paving materials.
- Enhancing urban planning to incorporate more vegetation and shade.
- Improving heat-resistant building designs.

8. **In 1953, the United States began using female names for naming storms, but this practice changed in 1979 when both male and female names were introduced. It wasn't until 2014 that the Met Office in the UK started giving storms male and female names in a similar tradition.**

The naming of storms, particularly tropical cyclones, serves several important purposes in meteorology. Prior to the 1950s, storms were often identified by their geographic location, which could be confusing and less effective for communication. The World Meteorological Organization (WMO) maintains lists of names for tropical cyclones in various ocean basins, often alternating between male and female and following alphabetical order. Names are typically retired if a storm is particularly destructive or deadly to avoid confusion and out of respect for the affected communities. This naming convention helps to raise awareness, facilitate communication, and enhance preparedness and response efforts for potentially hazardous weather events, contributing to the safety and resilience of communities worldwide. In Europe, storms are named through various national meteorological agencies rather than a centralised system like the one used for tropical cyclones. Countries across Europe have adopted their own methods for naming storms, often based on their historical practices and linguistic conventions. For example, the UK's Met Office and Ireland's Met Éireann jointly developed the "Name our Storms" campaign in 2015, where names are chosen from a list of submissions by the public.

9. **The Iran blizzard of February 1972 was the deadliest blizzard in history, as more than 7.9 metres (25 ft 11 in) of snow fell over a week and resulted in the deaths of over 4,000 people.**

   Blizzards are severe winter storms characterised by strong winds, heavy snowfall, and reduced visibility, posing significant risks to human life and infrastructure. These weather phenomena typically occur when cold air from the Arctic collides with warmer air masses, leading to intense snowfall and fierce winds. Among the deadliest blizzards in history is the Iran Blizzard of 1972, which struck Iran from 3rd February to 9th February 1972. This catastrophic event brought unprecedented snowfall and freezing temperatures, particularly affecting rural areas in northwestern and central Iran. The Iran Blizzard claimed the lives of an estimated 4,000 to 8,000 people, making it one of the deadliest blizzards in recorded history. Many victims were caught off guard as the heavy snowfall blocked roads and isolated entire villages, hindering rescue efforts and access to essential supplies. The wind and snow resulted in the breakage of trees and power lines; rails and many villages were buried, and vehicles were crushed beneath the weight of the snow.

10. **Flash floods are sudden and intense flooding events, often occurring within a short period, typically under six hours.**
These floods are commonly triggered by heavy rainfall, rapid snowmelt, dam or levee failures, or sudden release of water from ice jams. Flash floods can occur in various geographic settings, including urban areas, mountainous regions, and desert valleys. Due to their swift onset and powerful force, flash floods pose significant risks to human life, property, and infrastructure. Their ability to uproot trees, sweep away vehicles, and destroy buildings makes them among the most dangerous of all natural disasters. Vulnerable populations, such as those living in flood-prone areas or near riverbanks, are particularly at risk during flash flood events. Effective preparedness measures, including early warning systems, evacuation plans, and infrastructure improvements, are crucial for mitigating the impacts of flash floods and ensuring the safety and resilience of communities in at-risk areas.

11. **The polar vortex is a large area of low pressure and cold air surrounding both of Earth's poles. Occasionally, pieces of the polar vortex can break off and move south, causing frigid temperatures and extreme winter weather in regions that are typically not accustomed to such conditions.**

It typically resides near the North and South Poles in the Earth's upper atmosphere. Occasionally, parts of the polar vortex can weaken or break off, leading to the intrusion of frigid air masses into lower latitudes. These events are often associated with extreme cold weather outbreaks, particularly in regions such as North America, Europe, and Asia. The term "polar vortex" gained widespread attention in recent years due to its association with severe winter weather events, such as the polar vortex outbreak of January 2014, which brought record-breaking cold temperatures to parts of North America. While the polar vortex is a natural phenomenon, there is evidence to suggest that climate change may be affecting its behaviour, leading to more frequent disruptions and extreme weather events.

12. **Dust storms are created by strong winds that lift large amounts of dust and sand from dry, arid regions and carry them over long distances. These storms can reduce visibility to near-zero and have significant impacts on transportation, agriculture, and air quality.**
These storms typically occur in arid or semi-arid regions where loose soil and dry conditions create ample dust particles susceptible to being lifted into the air. Dust storms can be triggered by various factors, including thunderstorms, strong pressure gradients, and human activities such as land degradation and deforestation. They often result in reduced visibility, hazardous driving conditions, and respiratory problems due to the inhalation of fine dust particles. Dust storms can have significant socio-economic impacts, disrupting

transportation, damaging crops, and causing soil erosion. In regions prone to dust storms, such as parts of North Africa, the Middle East, and the southwestern United States, monitoring and early warning systems are essential for mitigating the risks associated with these natural hazards.

13. **Strong winds that cause dust storms can even carry dust across oceans and continents. For instance, dust from the Sahara Desert can travel vast distances across continents, carried by prevailing winds at various altitudes in the atmosphere.**

    As Saharan dust reaches other countries, it can affect the local environment and weather. One notable impact is the hazy or milky appearance of the sky, caused by the scattering of sunlight by the dust particles. This scattering effect can reduce visibility and create colourful sunrises and sunsets. High concentrations of Saharan dust can degrade air quality, leading to respiratory issues for sensitive individuals. However, Saharan dust also carries essential nutrients and minerals, which can enhance soil fertility and stimulate marine ecosystems' productivity when deposited onto land or into bodies of water.

14. **Weather modification is the deliberate alteration of atmospheric conditions to influence weather patterns. For example, for the 2008 Olympic Games in Beijing, the Chinese government invested heavily in cloud-seeding technology, deploying thousands of rockets**

containing silver iodide and other seeding agents into the atmosphere to ensure clear skies and reduce the risk of rainfall and pollution during the event by inducing rainfall before it reached the Olympic venues.

Various techniques are employed in weather modification, but one of the most common involves seeding clouds with substances like silver iodide, potassium iodide, or dry ice, which can enhance cloud formation and precipitation. Cloud seeding aims to increase rainfall, suppress hail, or disperse fog, among other objectives. While weather modification techniques have been explored for decades, their effectiveness remains a subject of debate and ongoing research. Some studies suggest that cloud seeding can modestly enhance precipitation in certain conditions, particularly in areas with suitable cloud structures, while others argue that the results are inconclusive or negligible. Concerns have also been expressed concerning the possible environmental and ethical implications of weather modification, including unintended ecological impacts and questions surrounding ownership and control of weather systems. Despite these challenges, weather modification continues to be studied and occasionally implemented for specific purposes, such as water resource management, agricultural enhancement, and hail suppression.

15. **Another example of weather modification by humans was Operation Popeye, a classified weather modification program conducted by the United States military during the Vietnam War**

in the late 1960s, with the goal of extending the monsoon season over targeted areas of the Ho Chi Minh Trail, a vital supply route used by North Vietnamese forces.
By cloud seeding with silver iodide, the military aimed to increase rainfall and create muddy conditions, thereby impeding enemy troop movements and supply deliveries. The operation, which ran from 1967 to 1972, involved aircraft flying over the target areas and releasing cloud-seeding agents into the atmosphere. While the exact extent of the operation's success is debated, some evidence suggests that it may have had localised effects on precipitation. Operation Popeye remains a controversial example of military weather modification efforts and has raised ethical concerns regarding the use of environmental manipulation as a weapon of war.

16. **Acid rain is a form of precipitation, such as rain, snow, or fog, that contains high levels of acidic compounds, mainly sulfuric acid and nitric acid.** These acids are created mainly by human activities, including burning fossil fuels, conducting industrial processes, and producing emissions from vehicles, releasing sulphur dioxide (SO_2) and nitrogen oxides (NO_x) into the atmosphere. Once released into the air, these pollutants react with water vapour, oxygen, and other chemicals to form sulfuric acid and nitric acid, which then fall to the Earth's surface during rainfall. The environment can suffer from acid rain in a number of ways, including acidification of soil and bodies of water, damage to vegetation and aquatic ecosystems,

corrosion of buildings and infrastructure, and adverse impacts on human health. Regulations to lower nitrogen oxide and sulphur dioxide emissions and global agreements to address cross-border pollution and promote cleaner energy technology have all been part of the effort to lessen acid rain.

17. **Human activity has significantly impacted weather patterns in various negative ways, primarily through the emission of greenhouse gases such as carbon dioxide, methane, and nitrous oxide.**
These emissions, which are mostly caused by the burning of fossil fuels for transportation, energy, and industrial processes, have led to global warming and climate change. The increased concentration of greenhouse gases in the atmosphere traps heat, leading to rising global temperatures, changes in precipitation patterns, an increased frequency and intensity of extreme weather events, and disturbances to ecosystems and agriculture. Deforestation, urbanisation, and land-use changes have further exacerbated these impacts, contributing to phenomena such as urban heat islands, loss of biodiversity, and degradation of natural habitats. Human-induced climate change poses significant risks to human health, livelihoods, and economies, highlighting the urgent need for mitigation and adaptation efforts to address these negative impacts and build resilience to future environmental challenges.

18. **Climate change is significantly impacting migration patterns worldwide as changing**

environmental conditions compel humans and animals to adapt to new circumstances.

Increased frequency of extreme weather events, changing precipitation patterns, and rising temperatures are disrupting ecosystems, decreasing the availability of resources such as food and water, and degrading habitats. In response, many species are shifting their ranges to more hospitable climates, while others face population declines or extinction. Similarly, human populations are also experiencing displacement and migration due to climate-related factors such as sea-level rise, desertification, crop failure, water scarcity, and increased frequency of natural disasters. Climate-induced migration can lead to social, economic, and political challenges, including competition for resources, loss of livelihoods, and conflicts over land and water.

19. **Lightning is a spectacular and powerful natural phenomenon that occurs when electrical charges build up within and between clouds and the ground. This buildup of electric charge creates an imbalance that seeks to be neutralised, resulting in the sudden discharge of electricity in the form of lightning. On Earth, on average, there are at least 8 million lightning strikes per day.**

    Lightning bolts can reach temperatures of up to 30,000 degrees Celsius (54,000 degrees Fahrenheit), hotter than the surface of the Sun, and can travel at speeds of over 200,000 miles per hour (320,000 kilometres per hour). Lightning strikes can take various forms, including cloud-to-ground, cloud-to-cloud, and intra-cloud lightning. They often produce brilliant flashes of light, followed by thunder, which is caused by the rapid expansion and contraction of air heated by the lightning bolt. Lightning poses significant risks to human safety, with an estimated 2,000 deaths worldwide by lightning strikes each year. However, lightning also plays a vital role in the Earth's atmospheric and ecological systems, as it helps to redistribute electrical charges, produce ozone, and sometimes initiate wildfires.

20. **We are currently living in the midst of an ice age. Known as the Quaternary Ice Age, it began approximately 2.6 million years ago and continues to this day.**
The Earth has experienced multiple ice ages throughout its history. They are characterised by alternating glacial and interglacial periods; glacial periods are long periods of cold climates where temperatures plummet, glaciers grow, and sea levels fall as water is locked up in ice. The last glacial period peaked about 20,000 years ago when much of North America, Europe, and Asia was covered in ice, sea levels were significantly lower, and animals like woolly mammoths roamed the Earth. Interglacial periods are warmer phases when glaciers retreat and sea levels rise. We are currently in an interglacial phase called the Holocene, which started around 11,700 years ago. The evidence of this ongoing ice age is visible in the polar ice caps and glaciers that still cover Greenland and Antarctica, as well as the remnants of glacial landscapes scattered across the globe. While the exact causes of ice ages are complex, factors such as variations in Earth's

orbit, changes in greenhouse gas levels, and continental drift are thought to play a role in initiating and ending these cycles.

21. **The 1936 North American heat wave was one of the most extreme and deadly weather events in recorded history, causing more than 5,000 deaths.**
Spanning June to August, the heatwave brought unprecedented temperatures across the United States and Canada, with some areas experiencing highs of up to 121°F (49.4°C). For several days, cities like Chicago and New York City saw temperatures soar above 100°F (38°C). The extreme heat was exacerbated by the ongoing drought, which had turned vast portions of the Midwest into a dust-laden wasteland. The combination of extreme heat, dust storms, and economic hardship during the Great Depression created a humanitarian crisis. Thousands of people lost their lives due to heat-related illnesses, and many more were forced to abandon their homes in search of relief.

22. **The Carrington Event in 1859 is the most powerful solar storm on record, named after British astronomer Richard Carrington, who observed the intense solar flare that triggered it. It created strong auroras that were reported globally and caused telegraph lines to spark, and gave some operators electric shocks.**
This extraordinary geomagnetic storm began when a massive coronal mass ejection (CME) from the Sun struck Earth's magnetic field, causing stunning auroras

visible as far south as the Caribbean and parts of Central America—locations where auroras are typically unheard of. In North America, people reported that the skies were so bright they could read newspapers at midnight by the aurora's light. The storm's impact on telegraph systems was especially dramatic; telegraph lines sparked, some operators received electric shocks, and in certain cases, telegraphs continued to transmit messages even when their batteries were disconnected, powered solely by the storm's intense geomagnetic currents. If a similar event occurred today, it could severely disrupt modern technology, knocking out satellites, power grids, GPS systems, and global communications, leading to potentially catastrophic economic and infrastructure impacts.

Thank you for reading!

Thank you for joining me on this journey through the wonders of the natural world. Nature has endless stories to tell; I hope these fascinating facts sparked curiosity and deepened your connection to the incredible planet we all share.

May this book be a starting point for many more adventures and explorations in the great outdoors. Thank you again for reading, and may you continue to find wonder in the world around you.

Happy reading!

Jemma Stone

www.ingramcontent.com/pod-product-compliance
Lightning Source LLC
Chambersburg PA
CBHW020424220526
45464CB00002B/560